A.

THOMA

Also by Johann P. Sommerville

POLITICS AND IDEOLOGY IN ENGLAND, 1603–1640

SIR ROBERT FILMER:
PATRIARCHA AND OTHER WRITINGS (*editor*)

Thomas Hobbes

Political Ideas in Historical Context

JOHANN P. SOMMERVILLE
Associate Professor of History
University of Wisconsin-Madison

© Johann P. Sommerville 1992

All rights reserved. No reproduction, copy or transmission of
this publication may be made without written permission.

No paragraph of this publication may be reproduced, copied or
transmitted save with written permission or in accordance with
the provisions of the Copyright, Designs and Patents Act 1988,
or under the terms of any licence permitting limited copying
issued by the Copyright Licensing Agency, 90 Tottenham
Court Road, London W1P 9HE.

Any person who does any unauthorised act in relation to this
publication may be liable to criminal prosecution and civil
claims for damages.

First published 1992 by
THE MACMILLAN PRESS LTD
Houndmills, Basingstoke, Hampshire RG21 2XS
and London

Companies and representatives
throughout the world

Copy-edited and typeset by Grahame & Grahame Editorial, Brighton

ISBN 0-333-49598-5 (hardcover)
ISBN 0-333-49599-3 (paperback)

A catalogue record for this book is available
from the British Library

Printed in Hong Kong

Contents

Preface vii

List of Abbreviations ix

Chronology xii

1 Hobbes and his Context 1
   1 Introductory 1
   2 1588–1640 5
   3 1640–51 19
   4 1651–79 23

2 The Law of Nature and the Natural Condition of Mankind 28
   1 Self-preservation 33
   2 The state and right of nature 37
   3 The laws of nature and justice 43
   4 Covenants 51

3 The Origins of Government and the Nature of Political Obligation 57
   1 The institution of a commonwealth 57
   2 Conquest 63
   3 The family and the state 70
   4 Political obligation 74

4 Hobbes on Sovereignty and Law 80
   1 Sovereignty 81
   2 The liberty and property of the subject 89
   3 Law and the sovereign 96
   4 The duties of the sovereign 100

5 Hobbes on Church and State 105
   1 What is Scripture? 108

	2	Ecclesiastical power: the case against Bellarmine	113
	3	Ecclesiastical power: *De Cive*, *Leviathan* and Anglican thinking	119
	4	Excommunication and Erastianism	127
6	God, Religion and Toleration		135
	1	God	137
	2	Faith, salvation, revelation, prophecy and miracles	141
	3	Toleration and conscience	149
	4	Church history and reformation	156

Conclusion	161
Notes	168
Bibliography	201
Index	222

Preface

This book is about the political ideas of Thomas Hobbes. Hobbes was a startlingly original thinker. I cannot say the same for myself. Many of my debts for particular points are recorded in the Notes and Bibliography. Tom Cogswell, Richard Cust, Sir Geoffrey Elton, Peter Lake, John Morrill, and Conrad Russell (Earl Russell) deserve thanks for answering my questions on Hobbes and his context, and more generally for guiding my thinking on early seventeenth-century English history. Mark Goldie, Quentin Skinner and Richard Tuck have heavily influenced my ideas on early modern political theory. All errors and misconceptions are, of course, my own.

The staffs of the British Library, Cambridge University Library, and Memorial Library here in Madison deserve thanks for their courtesy and efficiency. I am very grateful to the Graduate School of the University of Wisconsin at Madison, and especially to Jack Hexter and the John M. Olin Foundation, for funding my research and so enabling me to write this book. Patrick Riley deserves gratitude for making available to me his own set of Molesworth's edition of Hobbes' *English Works*, and for tactfully refraining from reminding me that I have not yet returned the books to him. I am very grateful to Quentin Skinner for sending me a copy of proofs of his pioneering essay on the relationship between Hobbes' political ideas and classical and Renaissance rhetorical theory ('Thomas Hobbes: Rhetoric and the Construction of Morality').

In the text below, dates are Old Style unless otherwise indicated, but the year is taken to begin on 1 January. I employ the term 'absolutist' 'to refer to accounts of political power which derive the ruler's authority either from a direct divine gift or an irreversible grant from the people', and which license the ruler to override all human laws in what he believes to be cases of necessity. This is Peter Lake's definition (Lake 1988, 7), and is (I think) how the word is generally used. 'Anglicans' are those who defended the government of the church of England as established after 1559. 'Laudians' are people who supported the policies of Archbishop William Laud; they commonly stressed the importance of sumptuous ceremonies in God's worship and rejected the idea that salvation and damnation are consequences of God's arbitrary decree. In the case of works which are available in many editions (e.g. Suarez's *De legibus*) references are generally given to the page numbers of a specified edition and also to book, chapter, and where appropriate section and subsection.

JOHANN P. SOMMERVILLE

List of Abbreviations

Aubrey	John Aubrey, *'Brief Lives,'* chiefly of Contemporaries, set down by Aubrey, between the Years 1669 & 1696, edited by Andrew Clark, 2 vols, Oxford 1898.
AW	Thomas Hobbes, *Thomas White's "De Mundo" examined*, translated by Harold Whitmore Jones, Bradford 1976. The original Latin was published in *Thomas Hobbes: Critique du "De Mundo" de Thomas White*, edited by Jean Jacquot and Harold Whitmore Jones, Paris 1973. References are to chapter and section, and to the folios of the original manuscript, which are printed in both the edition of the Latin text and in the English translation. Quotations are from the English translation.
Beh	Thomas Hobbes, *Behemoth or the Long Parliament*, edited by Ferdinand Tonnies, 1889; second edition, with an introduction by Stephen Holmes, Chicago 1990.
B.L.	British Library
CM	Cornelis de Waard and others, eds, *Correspondance du P. Marin Mersenne*, Paris, 17 vols, Paris 1932–88. Where appropriate, references are to the second, revised edition.
DC	Thomas Hobbes, *De Cive*. First published at Paris in 1642 under the title *Elementorum Philosophiae Sectio Tertia De Cive*. A revised and expanded edition appeared at Amsterdam in 1647, entitled *Elementa Philosophica De Cive*. The standard modern edition is *De Cive: the Latin version*, edited by Howard Warrender, Oxford 1983. A rather inaccurate English version appeared at London in 1651 under the title *Philosophicall Rudiments concerning Government and Society*, and has been edited by Howard Warrender as *De Cive: the English version*, Oxford 1983; this version has commonly been ascribed to Hobbes himself, but as

Richard Tuck and others have recently argued, there is strong evidence against this attribution: see e.g. Tuck, 'Warrender's *De Cive*', *Political Studies* 33 (1985), 308–15 at 310–12. References below are usually to chapter and section in Warrender's Latin version (e.g. DC2:6 refers to chapter 2, section 6), but occasionally to page number of editorial and supplementary materials (e.g. DC 60 refers to p. 60).

Dialogue — Thomas Hobbes, *A Dialogue between a Philosopher and a Student of the Common Laws of England*, edited by Joseph Cropsey, Chicago 1971.

EL — Thomas Hobbes, *The Elements of Law Natural and Politic*, edited by Ferdinand Tonnies 1889; second edition with a new introduction by M. M. Goldsmith, 1969. References are to part, chapter, and section.

EW — Thomas Hobbes, *The English Works of Thomas Hobbes*, edited by Sir William Molesworth, 11 vols, 1839–45.

HMC — Historical Manuscripts Commission Reports.

Lev — Thomas Hobbes, *Leviathan*, edited by Richard Tuck, Cambridge 1991. References are to chapter, page number in Tuck's edition, and page number in the 1651 edition. For example, Lev 21: 148/110 refers to chapter 21 of *Leviathan* at p. 148 of Tuck's edition, and p. 110 of the 1651 edition. The page numbers of the 1651 edition are printed in square brackets in Tuck's edition, and also in the widely used but less accurate edition of C. B. Macpherson (Harmondsworth 1968).

LW — Thomas Hobbes, *Opera Latina quae Latine scripsit omnia*, edited by William Molesworth, 5 vols, 1839–45.

SP — State Papers.

ST — Aquinas, St Thomas, *Summa theologiae*. References are to part, question and article.

STC — *A short-title catalogue of books printed in England, Scotland, & Ireland and of English books printed abroad 1475–1640*, ed. A. W. Pollard and G. R. Redgrave, revised by W. A. Jackson, F. S. Ferguson and Katharine F. Pantzer, 2 vols, 1976–86.

Wing   *Short-Title catalogue of books printed in England, Scotland, Ireland, Wales, and British America and of English books printed in other countries 1641–1700*, ed. Donald Wing, second edition, 3 vols, 1972–88.

Chronology

1586–9	Publication of Robert Bellarmine's *Disputationes de controversiis*
1588	Birth of Thomas Hobbes. Defeat of the Spanish Armada.
1603	Hobbes goes to Oxford University. James I becomes King of England.
1606	Pope Paul V places Venice under Interdict and condemns the English oath of allegiance (which rejected the claim that popes were empowered to depose heretical kings). Controversy over the Interdict and the oath.
1608	Hobbes graduates from Oxford and takes service with the Cavendish family as a tutor and later secretary.
1610–15	William Cavendish (later second Earl of Devonshire) tours the Continent, accompanied by Hobbes.
1612	Publication of Francisco Suarez's *De legibus ac Deo legislatore*.
1618	Outbreak of the Thirty Years' War.
1621	Debate between the House of Commons and James I on parliamentary privilege.
1625	Death of James I. Succession of Charles I. Publication of Hugo Grotius' *De jure belli ac pacis*.
1626–7	Charles I raises extra-parliamentary funds (the Forced Loan) which Hobbes helps to collect in Derbyshire.
1627	Sermons justifying the Forced Loan by Robert Sibthorp and Roger Maynwaring.
1628	Parliament approves the Petition of Right, condemning the Forced Loan and other recent royal policies.
1629	Publication of Hobbes' translation of Thucydides.
1629–30	Hobbes travels on the Continent with the son of Sir Gervase Clifton.
1629–40	Charles I rules without Parliament. Development of Hobbes' scientific interests.

Chronology

1634–6	Hobbes tours the Continent with the third Earl of Devonshire, and associates with Marin Mersenne and his circle at Paris.
1637	Publication of Hobbes' *A brief of the art of rhetorick*, summarising Aristotle's *Rhetoric*.
1637–8	Hampden's Case: the judges narrowly decide in favour of the legality of Ship Money.
1638–40	Scottish rebellion against Charles I.
1640	April: Charles I summons the Short Parliament to raise funds for war with the Scots. May: Charles I dissolves the Short Parliament, which had voted him no taxes. Hobbes completes the *Elements of Law natural and politic*. November: Under pressure from the victorious Scots, Charles I calls the Long Parliament, which begins constitutional reforms and attacks advocates of absolutist ideas. Hobbes flees for France.
1641	Execution of the Earl of Strafford, one of Charles I's leading supporters. Publication of Descartes' *Meditations on first philosophy*, including Hobbes' objections. Hobbes completes *De Cive*.
1642	Outbreak of Civil War in England. Publication at Paris of a small edition of *De Cive*.
1642–3	Hobbes writes a long reply to Thomas White's *De Mundo*.
1644	Royalist defeat at the battle of Marston Moor. William Cavendish (Marquis of Newcastle and a leading royalist general) flees for the Continent.
1645	Royalist defeat at the battle of Naseby.
1646	Civil War ends in defeat for royalists. Charles, Prince of Wales arrives at Paris, where Hobbes teaches him mathematics.
1647	Revised version of *De Cive* published in a large edition at Amsterdam.
1649	Execution of Charles I. Abolition of monarchy and the House of Lords. Establishment of a republic in England.

1650	Adult males required to take an Engagement, promising obedience to the new republic. Controversy over the Engagement. Cromwell invades Scotland.
1651	April/May: Publication at London of *Leviathan*.
	August: A Scots army invades England.
	September: Cromwell defeats Charles II and the Scots at the battle of Worcester.
1651/2	(Winter) Hobbes excluded from exiled court of Charles II in France. Hobbes returns to England and submits to the republican government.
1655	Publication of *De Corpore*.
1657	*Leviathan* reported to a parliamentary committee 'as a most poisonous piece of atheism'.
1658	Publication of *De Homine*.
1660	Restoration of Charles II.
1666	Parliamentary proceedings threatening to condemn Hobbes for his religious views. Hobbes writes *A Dialogue between a Philosopher and a Student of the Common Laws of England* (printed 1681)
1668	Revised Latin edition of *Leviathan* published at Amsterdam.
1670	Hobbes writes *Behemoth* (printed 1679)
1679	Death of Hobbes.
1683	*Leviathan* and *De Cive* condemned and burned at Oxford University.

1
Hobbes and his Context

1: INTRODUCTORY

This book – as its title suggests – is about the political ideas of Thomas Hobbes and his contemporaries. Hobbes' *Leviathan* is, of course, one of the greatest classics of political philosophy in the English language. It has much to say that is relevant to problems with which people have struggled through the ages, and which are still of the utmost concern today. It poses questions and offers answers on the nature of state power, and on the rights of women, minorities, and individuals. It is also, perhaps, the most stylishly written work of political theory in any language.

Hobbes' books are still extremely well worth reading today. Yet he wrote them not for us but for his contemporaries. He believed that they had made grave mistakes in their political thinking – mistakes which had resulted in the outbreak of civil war in England. His aim in *Leviathan* was to show them where they had gone wrong, and how they could construct a state which would bring lasting peace and harmony. If we are fully to understand what Hobbes said, it is well worth exploring the views of his contemporaries – for it was to their views that he was responding, and it was they whom he hoped to convince. The purpose of this book is to investigate the ideas of Hobbes in the context of the circumstances in which he formulated them, and of the theories which he opposed – or borrowed.

A contextualist approach to the classics of political theory has been championed by Quentin Skinner in a series of important essays.[1] Skinner has also written a number of excellent articles on Hobbes and his context. But though Skinner's approach to the history of political theory has proved highly influential, surprisingly few scholars have attempted to discuss Hobbes' political ideas in their historical context. This book tries to do just that. Since the book is about political ideas, and since space is limited, it has inevitably been necessary to say relatively little about many aspects of Hobbes' thinking – including his nominalism, his views on psychology and rhetoric, and the details of his scientific theories.[2]

Hobbes was certainly much influenced by the scientific ideas of his contemporaries. His political theory was intended to be a tight, quasi-geometrical system, in which conclusions flowed inevitably from premises. But, as his critic John Eachard pointed out, Hobbes' political conclusions were neither as novel nor as rigorously proved as he claimed. More recent commentators have made similar observations. According to Jean Hampton, *Leviathan* fails to provide 'a valid geometric deduction of Hobbes's political conclusions', while David Gauthier has noted that 'the defence of absolute sovereignty is an acknowledged weakness in Hobbes's argument'.[3]

This latter claim would, perhaps, have saddened Hobbes, since the main practical conclusion of his theory was precisely that no state can maintain security unless it is governed by an absolute sovereign. We will see below that many of the arguments Hobbes used to support this conclusion were close to those which his contemporaries employed. Hobbes frequently adopted or adapted familiar materials. His political theory was not (or not only) a deductive system, but (also) a dialogue with his contemporaries. As we shall see, one of his most characteristic techniques was to take commonly held views and, by introducing a few changes, employ them to reach unfamiliar conclusions. His analysis of the laws and state of nature, for instance, includes many traditional elements, but also some novel doctrines which permitted him to draw the extremely important (and unusual) practical conclusion that no one can hold rights of property against his sovereign – and therefore that kings may take their subjects' goods without consent. This principle was of the highest relevance to events of the 1630s, when Hobbes first formulated the main elements of his system, and when Charles I did indeed raise revenue without the consent of his subjects. Again, Hobbes used Scripture in very traditional ways to support the idea that popes have no power to interfere in the secular affairs of states. But he extended old notions in new directions to argue that the clergy hold no powers whatsoever (unless the sovereign chooses to grant them some) – a conclusion which annoyed churchmen.

This book is about the relationship between Hobbes' political ideas and other theories which circulated in his times. It argues that much of what he had to say becomes more comprehensible if we look at it in the context of the ideas and actions of his

1651 Leviathan

contemporaries. We shall see that the fundamental principles of Hobbes' political thinking remained largely unchanged between *The Elements of Law* in 1640 and *Leviathan* eleven years later – though *Leviathan* is far more outspoken on religious questions than his earlier works, and it sometimes refers to events of the Civil War period. The main context of his theory is to be sought in the intellectual climate in which he lived during the 1630s and 1640s. The two crucial elements here were royalism and the new science.

Such informed contemporary writers as the pseudonymous 'Eutactus Philodemius' and Sir Robert Filmer rightly regarded his theory as essentially royalist in character. Neither held that Hobbes had invented a wholly new variety of political thinking. In a careful survey of recent political writings published in 1658, the leading Presbyterian Edward Gee treated Hobbes as though he was taking part in just the same kind of theorising as thinkers like Suarez, Filmer, Grotius, Hooker and Ascham.[4] As we shall see, Hobbes frequently used arguments that others – and especially royalists – also employed. Indeed, the forty-second chapter of *Leviathan* (far and away the longest in the book) consists largely of such arguments. True, he loved paradox, and was fond of drawing unfamiliar conclusions from hackneyed premises.[5] But even here he was using novel moves in an old game, not creating a new one; the game was the late-sixteenth and early-seventeenth-century European-wide debate on the nature and limitations of royal power. Hobbes' thinking was virtually unaffected by the advent of Leveller radicalism in the 1640s, or by Harringtonian republicanism (which acquired some vogue in England in the later 1650s).

Commenting on Hobbes' *De Cive* of 1642, the great French philosopher Descartes remarked that its author's whole goal was to write in favour of monarchy, but that this could have been better done by adopting more solid and virtuous principles than those upon which he in fact grounded his case.[6] Hobbes attempted to base his theory on a single principle – that of self-preservation. Earlier writers had indeed placed a good deal of weight upon this principle, but none had tried so single-mindedly to deduce a whole system from it.[7] As we shall see in Chapter 2 below, many of the oddities (in contemporary eyes) of Hobbes' moral system result from the fact that he tried to deduce his conclusions from this one principle. But we shall also see that in many respects his moral thinking was not so far removed from that of his contemporaries.

In Chapter 3 we shall move from Hobbes' general moral theory to his account of the origins of government. Though Hobbes presents his argument as a series of deductions from first principles, he appeals along the way to the debates and concepts of his contemporaries. For instance, his discussion of the institution of commonwealths is steeped in the casuistry of covenanting, and his treatment of conquest and the family is shot through with allusions to current thinking. Having surveyed how governments come into being, Hobbes proceeded to discuss their powers – and this is the subject of Chapter 4. Here Hobbes differed from most royalists for he denied that subjects have rights of property against their sovereign, and that the sovereign can ever commit an act of injustice against the subject. But on many questions about sovereign power he was in substantial agreement with other royalists – and the significance of the divergences should not be overestimated.

A great deal of *Leviathan* is concerned with church-state relations and religion – though these topics have been notoriously neglected by most of those who have written on Hobbes. As we shall see in Chapter 5, Hobbes employs traditional Anglican anti-papal arguments on church-state relations but turns them to criticise not only Catholic but also Presbyterian and even Anglican attitudes. The technique is reminiscent of William Chillingworth, whose *The religion of Protestants a safe way to salvation* (1638) was ostensibly an attack on Roman Catholicism but also put forward principles that could easily be used to criticise many orthodox Anglican notions. According to Aubrey, Hobbes used to say that Chillingworth 'was like a lusty fighting fellow that did drive his enemies before him, but would often give his own party smart back-blows' (1:173; cf. 1:371). Chillingworth and other members of Tew Circle held religious views which resembled those of Hobbes on a number of points, and it is sometimes argued nowadays that Hobbes was an orthodox latitudinarian Anglican in religion. As we shall see in Chapter 6, Hobbes differed on some crucial questions from the latitudinarians and this helps to explain why in the years after the Restoration he continued to attract charges of atheism while latitudinarians acquired increasing influence within the church of England. But before we turn to these topics, it will be useful to say something about his life and times. That is the subject of this chapter.

2: 1588–1640

Thomas Hobbes was born in 1588, the year of the Spanish Armada. He later said that it was fear of the Spaniards which led his mother to give birth to him prematurely, and he described himself and fear as twins (Latin verse life, in LW1:lxxxvi). Hobbes' political theory gave a special place to the emotion of fear, and he liked to stress his own fearfulness – though his life and writings show that he in fact possessed considerable courage. His birthplace was Westport (near Malmesbury in Wiltshire). Hobbes was the second son of a minor cleric, also named Thomas. This elder Thomas was not a model clergyman. He lacked any great academic knowledge, and, according to Hobbes' biographer Aubrey, he 'disesteemed learning' (Aubrey 1:323). On one occasion, says Aubrey, he spent Saturday night playing cards and then fell asleep in church on Sunday. While sleeping he announced that 'trafells' (i.e. clubs) were trumps. Later he hit a fellow-clergyman during an argument, and then fled from the area (Aubrey 1:387).[8]

Hobbes' father was an undistinguished clergyman. But the elder Thomas had a brother named Francis who was a childless and wealthy glove-manufacturer. Francis financed young Thomas' university education (Aubrey 1:324). This took place at Magdalen Hall in Oxford from 1603 to 1608, when Hobbes graduated as a bachelor of arts (Aubrey 1:330). Even before going to Oxford, he already displayed considerable academic promise, and translated Euripides' *Medea* from Greek into Latin verse (Aubrey 1:328–9). At Oxford he learned scholastic logic and physics, and also spent time looking at maps in local book-shops. In the summer he was fond of getting up early to bait jackdaws. (Aubrey 1:329; Hobbes' prose life in LW1:xiii; verse life in LW1:lxxxvi–ii).

The year of Hobbes' birth coincided with the sailing and defeat of the Spanish Armada, which the Catholic Philip II of Spain had mounted in order to oust the Protestant Elizabeth I from the throne of England. In 1570, Pope Pius V had issued a bull declaring that Elizabeth was excommunicated and deposed. Catholics commonly claimed that the pope is the spiritual overlord of all Christians, and that because spiritual matters are more important than temporal ones he is empowered to interfere in temporal affairs – for example by deposing monarchs – if he thinks this is in the spiritual interest of Christians. Hobbes' Oxford years coincided with the beginning of a European-wide controversy on the question of the pope's claim that he could depose kings. In 1606 the English Parliament passed a statute imposing stringent penalties on Catholics who refused to take an oath of allegiance abjuring the papal deposing power. In the year of Hobbes' graduation, King James I published a book defending the oath and attacking the deposing power. The king and his supporters soon became embroiled in a literary war with Catholic theologians, including such notable figures as Francisco Suarez and Robert Bellarmine.[9] The forty-second chapter of *Leviathan* makes it clear that Hobbes was familiar with the main arguments used by both sides in this debate. Hobbes himself devoted many pages to refuting the claims of Bellarmine (see below, Chapter 5, section 2).

A common choice of career for a graduate of Hobbes' background was the church. But a different kind of opportunity presented itself when in 1608 he was invited to join the household of the wealthy William Cavendish as tutor to his son and heir – also named William. From then on Hobbes was to be closely connected with the Cavendish family until his death in 1679. He never married, and Aubrey records that 'he was, even in his youth, (generally) temperate, both as to wine and women' (1:350).[10] In 1605 the elder William had been elevated to the peerage as Baron Cavendish of Hardwick. In 1618 he became Earl of Devonshire. The younger William inherited the earldom on his father's death in 1626, but held it for only two years, dying in 1628. His son – another William – then became third Earl of Devonshire. Like his father, he was tutored by Hobbes, who also served both men as secretary and friend. The first Earl had a younger brother named Charles Cavendish, and this Charles had two sons who were both patrons of Hobbes. The elder son – yet another William – became Earl, then Marquis and finally Duke of Newcastle. In 1638 Newcastle was appointed governor of the Prince of Wales, and during the Civil War he was one of Charles I's most important supporters, using his vast wealth to raise an army for the king. He took a keen interest in Hobbes' work, and it was to him that the philosopher dedicated his first political treatise, the *Elements of Law*. Newcastle's brother Sir Charles Cavendish was a patron of scientists and himself an able mathematician.[11]

Between 1610 and 1615 William, the future second Earl of Devonshire, toured the Continent, and Hobbes went with him. In Venice, William struck up a friendship with Fulgenzio Micanzio, a close associate of the leading Venetian statesman and intellectual Paolo Sarpi. Venice had been put under Interdict by Pope Paul V in 1606 because of its opposition to papal claims. The Interdict – which prohibited the performance of religious rites – resulted in a pamphlet debate which covered much the same ground as the dispute on the English oath of allegiance. Sarpi himself wrote against papal pretensions, and eagerly read books on the subject published by English presses. In 1612 James I invited him to England. Although professedly Catholics, Micanzio and Sarpi shared much of the outlook of James and his English supporters on questions of church-state relations. Like James, the two Venetians vigorously rejected the papal deposing power. Venetian and English anti-papalists employed many of the same arguments in these controversies. Their works were mostly written in simple scholastic Latin (that is to say, the Latin of medieval universities) and avoided the rhetorical flourishes commonly associated with humanism – though a few (including Bishop Lancelot Andrewes and his friend the great classical scholar Isaac Casaubon) did opt for an ornate humanist style. The essence of their case was that it is possible to read off from human nature a binding moral law (the law of nature) and that this law guarantees the autonomy of the state in secular matters. Since God created human nature, the argument proceeded, the law of nature has been willed by him and must be compatible with his will as recorded in the bible. Nothing that the bible said, therefore, could destroy the state's autonomy. So the papalist argument that Christ had granted the pope power over kings in temporal matters – and subjected individual states to the Catholic church – was based on misinterpretations of Scripture. The whole of this argument was derived from medieval scholastic writers.

When William Cavendish returned to England in 1615 he soon entered into a correspondence with Micanzio. Hobbes translated the latter's letters from Italian into English and they circulated amongst the earl's friends. Sarpi and Micanzio were keen to ensure that England took a vigorously anti-Spanish stance in international politics, since they believed that Spain was bent on a course of aggressive expansionism which threatened Venice and which was to be carried out under the cloak of a militant and papalist Catholicism. In 1618 the Thirty Years' War broke out on the Continent and for a time it looked as though the Spaniards would indeed destroy their opponents. Many people expected that England, the most powerful Protestant kingdom in Europe, would enter the war against Spain. But James I instead pursued a policy of peace and dynastic alliance with the Spanish crown.

Micanzio's letters to Cavendish record his dismay at this. He also had much to say about a fellow-Venetian, Marc' Antonio De Dominis, who came to England in 1616 and published voluminously against the papacy and its defenders – notably Suarez and Bellarmine. In *Leviathan* Hobbes employed many of the arguments against the pope's powers which had featured in the work of De Dominis. At first Micanzio was on friendly terms with his countryman, but this changed during the latter's stay in England. De Dominis took the side of the Arminians (who argued that human free will plays a role in salvation) in their disputes with predestinarian Calvinists (who claimed that salvation and damnation are consequences of God's arbitrary decrees). The controversy had profound effects both in the Netherlands and in England. Micanzio and Sarpi feared that Arminianism would undermine the Protestant churches and so promote the Spanish cause. De Dominis did indeed return to the church of Rome in 1622 – an event which Hobbes reported to his friend Robert Mason, a Cambridge academic. Later, Hobbes was vigorously to reject Arminian ideas on free will, though on many questions of political theory his views were close to those of the English Arminians and De Dominis.[12]

De Dominis' anti-papalist writings were scholastic in form and content. This does not mean that he was uninterested in new scientific developments. While in England he translated Bacon's *De sapientia veterum* into Italian. Earlier, he had published a book on optics – a subject on which Hobbes was also to write. Micanzio, too, had scientific interests. He greatly admired Bacon, whose writings he frequently discussed in his letters to Cavendish – a friend of Bacon's. In 1616 Cavendish told Bacon about Micanzio's interest in his work, and a correspondence began between the two men. Later, Micanzio was in close contact with another of the greatest scientific thinkers of the age, Galileo.[13]

Hobbes himself sometimes acted as secretary to Bacon and translated a number of his *Essays* into Latin (Aubrey 1:331). Bacon stressed the importance of experiment in acquiring scientific knowledge, while Hobbes was to lay more emphasis upon the need for rigorous argument from first principles. But both men believed that Aristotle's scientific system, which was still taught in the universities, was so flawed that it needed to be abandoned altogether. Both, indeed, planned to construct a new system of ideas to replace Aristotelianism. Again, both inveighed against the use of jargon in philosophical writings and stressed the need for clear and precise language.

After he came back from touring the Continent with William Cavendish, Hobbes spent much time studying the Greek and Roman classics. His writings display knowledge of a very wide range of ancient authors. He translated Thucydides' history of the Peloponnesian war from Greek into English and by November 1628 the translation was in the press.[14] Later Hobbes claimed that his purpose in publishing this book was to show his fellow citizens the folly of democracy and to warn them against listening to the demagogic orators (Latin verse life in LW1:lxxxviii; Latin prose life in LW1:xiv).

There were good reasons for thinking this message particularly apposite in 1628. As we saw, James I was reluctant to go to war with Spain and instead negotiated for an alliance, including the marriage of his son and heir Charles to a Spanish princess. But in 1624 the marriage negotiations finally broke down, and Charles – in conjunction with the king's favourite George Villiers (Duke of Buckingham) and many in parliament – now pressed for a war. When James died in 1625 Charles succeeded him and sent forces against Spain. Not long afterwards, England also went to war with France. Military operations were underfinanced and mismanaged, and many blamed Buckingham for this. To prevent the parliament of 1626 from impeaching the Duke, Charles dissolved it before it had voted him any money. Needing funds for the war, the king now resorted to the unorthodox step of raising the so-called 'Forced Loan' in 1626–7, coercing his subjects into 'lending' him money with very doubtful prospects of repayment. There was widespread ill-feeling at this measure. Many people believed that the monarch had no right to take property without the consent of parliament.

Hobbes – as secretary to the Earl of Devonshire – helped to collect the Forced Loan, and to send payments to the Exchequer.[15] In order to persuade people that they should pay the Loan, the king instructed the clergy to preach in its favour. Two clerics – Roger Maynwaring and Robert Sibthorp – responded to this call with particular enthusiasm. Their sermons, which were soon printed, argued that the king derived his power from God alone and that he was not accountable to the people. If the monarch thought it necessary, they claimed, he was perfectly entitled to demand money from his subjects, and the latter should pay it cheerfully. In their opinion, ideas of limited monarchy and legitimate resistance were characteristic of Jesuits and puritans, and ought to be rejected by all true-hearted Protestants. Sibthorp vigorously attacked Catholics for placing 'the church above the king, and the pope above the church', and puritans ('that factious fraternity') for setting 'the law above the king, and the people above the law'.[16] Maynwaring drew on the works of De Dominis, Andrewes and similar writers to vindicate Charles' right to levy the Loan. Aubrey later recorded that Hobbes 'told me that Bishop Maynwaring (of St David's) preached his [i.e. Hobbes'] doctrine' (Aubrey 1:334).

In addition to taking money without parliamentary consent, Charles also carried out a number of other unpopular actions in 1626–7 – billeting troops in civilian homes, for instance, and imprisoning refusers of the Loan without bringing any charge against them. Eventually, the king was persuaded to abandon these policies, and to call another parliament. This met in 1628 and members soon began to attack recent royal acts. The House of Commons approved a Petition of Right which declared that taxation without parliamentary consent was illegal and which condemned many of the other things that Charles had been doing. Imprisonment, it said, should take place only according to 'the law of the land' and not at the king's discretion.[17] The Commons also began impeachment proceedings against Maynwaring.

In the Lords there were considerable misgivings about the Petition, and the second Earl of Devonshire was one of many who objected to the clause about imprisonment. Charles had called parliament because he wanted it to vote him money. The Commons were unlikely to grant the king cash unless he approved the Petition. In the end, the Lords dropped their objections to the document. They also sentenced Maynwaring to pay a fine of a thousand pounds, and to be forever disabled from holding any 'ecclesiastical dignity or secular office'. Charles approved the Petition, took the money, and promptly pardoned and rewarded Maynwaring, who became Bishop of St David's a few years later. After encountering further hostility towards his policies from members of the Commons, the king dissolved parliament in March 1629 and issued a declaration vigorously attacking 'some ill affected persons of the House of Commons' by whose 'seditious carriage' 'Wee, and our Regall authority and commandement, have beene so highly contemned, as Our Kingly Office cannot bear, nor any former age can parallel'. Charles made it plain that he had no intention of calling another parliament in the near future, and did not in fact do so until 1640.[18]

Hobbes' position in the Cavendish household gave him opportunities to meet a great many English intellectuals and men of letters. The family had investments in the Virginia Company, and Hobbes himself became a member in 1622. He sometimes attended meetings of the Company between that date and its dissolution in 1624. There he rubbed shoulders with such notables as Sir Edwin Sandys, the poet John Donne, and (perhaps) the great scholar and lawyer John Selden.[19] His literary acquaintances also included the poet and dramatist Ben Jonson, and Sir Robert Ayton, a Scottish poet who was related to the second Earl of Devonshire's wife Christiana. Hobbes asked these two to comment on the style of his translation of Thucydides (Aubrey 1:365).

Aubrey tells us that George Aglionby (or Eglionby) was also Hobbes' 'great acquaintance' (Aubrey 1:370). Aglionby was a clergyman who took the degree of Doctor of Divinity in 1634, and who for a while was tutor to the second Duke of Buckingham (the son of James I's favourite). In 1643 he was appointed to the deanery of Canterbury. Shortly afterwards he died at Oxford, the royalist headquarters.[20] There is a letter to Hobbes from an Aglionby, dated 1629. In June 1628 the second Earl of Devonshire died. During 1629 Hobbes once more set out for the Continent, this time in the company of the son of Sir Gervase Clifton (a neighbour of the Earl of Newcastle). In November 1629 Aglionby wrote to Hobbes relaying recent political news, including proceedings against various notables for circulating a manuscript treatise entitled 'The present policy and government of the state and church of England'. Their purpose in doing so was reputedly 'to advance the liberty and defeat the prerogative'. Aglionby displayed little sympathy for the cause of liberty, and made plain his hostility to puritans and common lawyers. He clearly believed that Hobbes shared his attitudes.[21]

George Aglionby was a friend of Lucius Cary, Viscount Falkland (Aubrey 1:151) who in the 1630s famously played host to a group of intellectuals at his house at Great Tew not far from Oxford. Aubrey records that Falkland himself was a 'great friend of Hobbes' (Aubrey 1:365, cf. 1:151), and amongst Hobbes' other acquaintances he lists Falkland's close associate William Chillingworth (1:370). Edward Hyde, who became Earl of Clarendon and served Charles II as Lord Chancellor, was a particularly good friend of Falkland. Many years later Clarendon said that Hobbes 'is one of the most ancient acquaintance I have in the world'. The members of Falkland's group – known as the Tew Circle – did not see eye to eye on all matters, but most of them adopted a latitudinarian approach to religious questions and became royalists in the Civil War. As we shall see, some but by no means all of the religious opinions voiced by Falkland, Chillingworth, John Hales and Jeremy Taylor were strikingly similar to positions adopted by Hobbes. Again, a number of political arguments employed by the pamphleteer Dudley Digges in the early 1640s were unusually close to some of Hobbes' key tenets.[22]

It was, Hobbes later said, during his Continental tour with young Clifton that he first began to take an interest in geometry, and especially in Euclid's *Elements* (Latin prose life, in LW1:xiv). The demonstrative certainty of geometrical theorems greatly impressed him (Aubrey 1:332), though it is unclear quite how far his mathematical and scientific ideas had developed by 1630.[23] We may be sure that Hobbes' scientific and philosophical knowledge grew in the 1630s, though the exact chronology of this is obscure. In November of 1630 he was back at the Cavendish home in Derbyshire, and he now became tutor to the third Earl of Devonshire.[24] During the next few years he came to be increasingly closely associated with a group of scientists who enjoyed the patronage of Newcastle at Welbeck Abbey. Early in 1634 Hobbes was in London trying to find a copy of Galileo's highly important *Dialogue concerning the two chief world systems* (1632) for Newcastle.[25] In addition to Newcastle and his brother Sir Charles, the Welbeck group included Robert Payne (officially engaged as Newcastle's chaplain, though in fact employed largely to do scientific work), and it was in close touch with other scientists, including the noted mathematicians Walter Warner and John Pell. Warner, who was a former associate of the eminent scientist Thomas Harriot, lived in the household of Sir Thomas Aylesbury, the Master of the Mint and a keen patron of mathematical learning. Aylesbury was the father-in-law of Edward Hyde, and it is certain that Payne and Hyde were acquainted in the 1630s. Warner wrote to Payne in October 1634, mentioning a problem connected with refraction, and requesting Payne 'by any means' to 'send it to Mr Hobbes, together with my most hearty love and service'. 'I have found him free with me', said Warner, 'and I will not be reserved with him, if it please God I may live to see him again'.[26] This letter demonstrates that Hobbes was already gaining a reputation in the field of optics by 1634.

In the same year Hobbes returned to the Continent yet again, this time in the company of his pupil the third Earl of Devonshire. There he met Galileo, and also a number of French mathematicians and scientists, including the Minim friar Marin Mersenne. Mersenne corresponded with scientists in a number of countries, relaying the latest information about publications and about research in progress (a parallel figure in England was Samuel Hartlib). His associates included the greatest of French philosophers, René Descartes, and Pierre Gassendi – who attempted to revive the ideas of the ancient Epicureans, and who became a close friend of Hobbes in the 1640s.

While on the Continent, Hobbes stayed in touch with friends in England. He asked one to send him recent English publications on the sabbath – a topic that was hotly debated in the later 1630s (EW7:454). He wrote to Sir Charles Cavendish on optics,[27] and to Newcastle, discussing a variety of topics, including physical experiments and horses. In August 1635 he announced his intention of being the first writer to speak sense in plain English on the 'faculties and passions of the soul'[28] – indicating that he was already at work on the philosophy of human nature, and psychology. In June 1636 he reported that he was reading an important work of political theory, John Selden's recently published *Mare Clausum* – a defence of the English crown's claim to dominion over the seas, written in response to the *Mare liberum* of the great Dutch thinker Hugo Grotius.

Hobbes was also studying 'a book of my Lord of Castle Island's concerning truth, which is a high point'.[29] This was Edward Lord Herbert of Cherbury's *De veritate*, which had first been printed at Paris in 1624. The great French philosopher Descartes discussed the book with both Hartlib and Mersenne – who may have been the author of a French translation of it, published in 1639. Like Herbert, Mersenne and Descartes were both very concerned with finding responses to the arguments of ancient and modern sceptics, whose claims seemed to undermine any possibility of knowledge. A problem with Herbert's work was the obscurity of its prose style. In October 1636 Robert Payne reported that he and others found the book incomprehensible.[30]

That same month, Hobbes was back in England once more. Shortly before he left France he had been in contact with another correspondent of Mersenne's, Sir Kenelm Digby.[31] Digby was a great admirer of 'that never enough praised gentleman, Monsieur Descartes'. He also much admired Hobbes, whom he addressed in a letter of 1637 as 'You that know more than all men living'. This letter strongly suggests that Digby already recognised Hobbes as a major philosophical figure – and he was himself no mean philosopher. In

October 1637 Sir Kenelm sent Hobbes a copy of Descartes' recently published philosophical classic, the *Discours de la méthode pour bien conduire sa raison, et chercher la vérité dans les sciences* (Discourse on the method of correctly conducting one's reason and seeking truth in the sciences). Descartes, like Hobbes, set great store by the methods of geometry. His book contained an essay on optics in which he argued that colour (like other secondary qualities) exists not in the object of vision but in the senses of the beholder, and that it is caused by movements affecting the nerves.[32] Later, Hobbes was to be much concerned with showing that he had himself expressed the same viewpoint as early as 1630, and that he had not cribbed it from the Frenchman.[33]

Early in 1640 Hobbes completed his first political treatise, the *Elements of Law, Natural and Politic*. The epistle dedicatory, which was dated 9 May 1640 and addressed to the Earl of Newcastle, set out his reasons for writing the book. 'From the two principal parts of our nature, Reason and Passion', he said, 'have proceeded two kinds of learning, mathematical and dogmatical'. The former 'consisteth in comparing figures and motion only; in which things truth and the interest of men oppose not each other'. The latter, by contrast, concerned people's 'right and profit; in which, as oft as reason is against a man, so oft will a man be against reason'. On mathematical questions, people were willing to listen to reason. But on questions of 'justice and policy' they followed their emotions – for these questions involved the individual's self-interest. To write accurately and persuasively on ethics and politics it was therefore necessary to begin with principles which were so certain that no one, however swayed by passion, could doubt them. From such first principles, conclusions might then be drawn by rigorous logic. Hobbes held that the *Elements of Law* achieved precisely this, presenting the work to the earl as 'the true and only foundation of such science' (EL, epistle dedicatory, xv–xvi).

The outlook which informed this (by no means overly modest) dedicatory letter owed much to attitudes common in Mersenne's circle. Gassendi, in the first book of his *Exercitationes paradoxiae adversus Aristoteleos* (Paradoxical dissertations against the Aristotelians) (1624), referred to Aristotle's followers as the 'dogmatic philosophers' and inveighed against them (amongst other reasons) for their hostility to mathematics and their methodological confusions. Descartes mounted what he took to be rigorous arguments from self-evident first principles in his *Discours de la méthode* and elsewhere. He tried to show that a great many truths about the world could be generated from a few obvious premises – and in particular from the indubitable truth that 'I think, therefore I am' (*cogito ergo sum*). Hobbes similarly attempted to generate an extensive political philosophy from the single axiom of self-preservation. When Hobbes anonymously published the second version of his political theory – *De Cive* – in 1642, the French intellectual Samuel Sorbière at first suspected that the book was by Descartes – though René soon disabused him, declaring that 'he would never publish anything on morals'.[34] In form and method, then, the *Elements* bore the stamp of Mersenne's group. But in content – as we shall see – it owed as much to earlier absolutist thinking, and to the political debates of the 1620s and 30s.[35]

We saw that Sibthorp and Maynwaring claimed that monarchs derive their powers from God alone, and that they are not accountable to their subjects. Such views were commonplace in the higher ranks of the English clergy. According to the controversial canons of 1640, drawn up as an official statement of the clerical viewpoint, 'The most High and Sacred order of Kings is of Divine right, being the ordinance of God himself, founded in the prime Laws of nature, and clearly established by expresse texts both of the old and new Testaments'.[36] Similar notions were famously expressed by James I and by Sir Robert Filmer. The theory is commonly known as 'the divine right of kings', and is sometimes rather strongly contrasted with Hobbes' doctrines.[37] As we shall see, Hobbes' position did indeed differ from that of most contemporary apologists for absolute royal power, but it does not do to exaggerate the differences.

It is true that Hobbes' self-consciously refrained from citing authorities, and constructed his political arguments as rational deductions from first principles. Like many in the Mersenne circle, and like Bacon, he found much to criticise in the writings of Aristotle and the scholastics. But few political writers were blind followers of Aristotle, and most claimed that the principles they put forward were derived from reason not authority. Questions of political rights and duties were standardly held to be answerable by reference to reason, which granted people access to the laws of nature. This was what the canons of 1640 was getting at when they said that kingship was 'founded in the prime Laws of nature'. These laws were seen as unalterable moral principles, discoverable by reason. The laws of nature, it was argued, were compatible with the will of God expressed in the bible. Indeed, it was said, many of the natural laws were specifically confirmed and given lucid formulation in Scripture. But the laws themselves were held to apply to non-Christians as well as Christians, for everyone possessed reason. Many writers quoted the bible to support their rational arguments on political rights and obligations. Hobbes did the same.

Proponents of the divine right of kings rarely argued that monarchy was the only valid form of government. Rather, their point was that all societies need to be governed, and that if government is to be stable the governor(s) must ultimately be unaccountable to his (or their) subjects. True, they claimed that monarchy was normally the best kind of government, but few argued that this arrangement was unalterable. The usual doctrine was that God can intervene in human affairs to change the ruling dynasty, or to replace monarchy with aristocracy or democracy. An implication of this was that hereditary right is *not* indefeasible. Supporters of divine right monarchy placed particular emphasis upon refuting arguments in favour of the ideas that subjects can sometimes resist their sovereigns, and that the power of monarchs is limited by contractual obligations imposed upon them by the people. Hobbes likewise argued strongly against notions of limited monarchy and legitimate resistance, and claimed that kingship was the best though not the only valid form of government.

Hobbes was later to assert that 'civil philosophy' is 'no older than my own book *De Cive*' (EW1:ix; cf. LW1:cv). In the epistle dedicatory of that work (addressed to the third Earl of Devonshire) he argued that modern progress in knowledge was wholly due to geometry – for physics itself depended on geometry (DC epistle dedicatory, 5). Moral philosophers had failed to rival the achievements of geometers, he held, because they had not reasoned properly from first principles. He himself had, and this was why he alone had created a science of politics (DC epistle dedicatory, 6–10).

Though boastful, there was some truth in Hobbes' claims. Much contemporary political writing took the form of works of controversy intended to refute the specific arguments of an opponent and not to put forward any comprehensive philosophical position. Of course, some authors did compose more systematic treatises. Richard Hooker's *Laws of ecclesiastical polity* is one example (though it was in part a work of polemic). Still clearer instances are the Spaniard Francisco Suarez's *De legibus* (On Laws; 1612), and the Dutchman Hugo Grotius' *De jure belli ac pacis* (On the law of war and peace; 1625). Hobbes was certainly familiar with at least some of the works of Suarez (Lev 8:59/39; EW5:37) and it is highly likely that he had read Grotius carefully. Both Suarez and Grotius attempted to provide closely reasoned, exhaustive and systematic treatments of their subject. But neither adopted the geometrical method which, in Hobbes's view, was alone suited to establishing a *science* of politics. So to that extent Hobbes' claims were justified. Both Grotius and Suarez did, however, mount many arguments which recur in Hobbes. Frequently, too, he went out of his way to refute positions which such writers had maintained. Hobbes' biographer Robertson plausibly argued that much of what he had to say was directed against Grotius, even though Hobbes did not mention the Dutchman.[38] Hobbes meant his political writings to be not only abstract quasi-geometrical treatises but also vital contributions to contemporary debate.

In the years before the Civil War, political debate in England focussed largely on the questions of the nature and limitations of royal power. After dissolving parliament in 1629, Charles I once again reverted to some of the policies which had annoyed so many of his subjects in the 1620s. He imprisoned people without showing legal cause, and raised funds without parliamentary consent. The most famous of these levies was Ship Money, which the king used to improve the English navy. In John Hampden's Case of 1637–8, the judges decided by the slimmest possible majority that the levy was legal. But when circumstances once more compelled Charles to summon parliament in 1640, it soon became evident that a great many people believed that this decision was unlawful and that Ship Money was contrary to the Petition of Right and to the fundamental liberties of Englishmen. Henry Parker, one of the most famous of parliamentarian pamphleteers, vigorously condemned Ship Money in a pamphlet of 1640, inveighing against theories of royal absolutism and its advocates, who were, he said, 'the Papists, the Prelates, and Court Parasites'. It was 'the common Court doctrine', he declared, 'that Kings are boundlesse in authority'.[39]

By 1640 the Earl of Newcastle was an important figure at Charles' court. The dedicatory epistle to him, prefixed to the *Elements of Law*, strongly suggests that Hobbes expected the earl to agree with his views and indeed to pass the book on to 'those whom the matter it containeth most nearly concerneth' – presumably the king himself or his closest advisers, such as Archbishop Laud and the Earl of Strafford (EL xv–xvi). In the years between 1629 and 1640 a number of clerics had followed in the wake of Maynwaring by expressing outspokenly absolutist opinions – and with no parliament sitting it had been comparatively safe to do so. Those who did included Archbishop William Laud, Peter Heylin (the Archbishop's friend and biographer), and William Beale. Amongst the laity, the courtier Sir Francis Kynaston voiced similar ideas in a manuscript treatise.

So too did Sir Robert Filmer, who had court connections and was a friend of Heylin's. Filmer had completed a manuscript of his famous *Patriarcha* by 1632. In the 1640s and 1650s he published several political pamphlets trenchantly asserting absolutist theory. During those years many royalist pamphleteers found it prudent to tone down or abandon absolutist thinking in order to emphasise the moderation of the royalist cause. But Filmer stuck by his old principles. In his later political writings Hobbes likewise made no effort to disguise his absolutism – and indeed clearly enjoyed shocking his readers by spelling out his conclusions. The two men shared very similar opinions on the powers of rulers and the duties of subjects – opinions which they had come to develop by the 1630s at the latest. Commenting on *De Cive* and *Leviathan* Filmer said that no one had discussed 'the rights of sovereignty' as 'amply and judiciously' as Hobbes.[40]

In the later 1630s Charles I attempted to impose a new prayer book upon the Scots, who responded by rebelling against him. In 1639 Newcastle raised troops at his own expense to fight the rebels. Needing money for the war, Charles called parliament in April 1640. The Earl of Devonshire unsuccessfully attempted to have Hobbes elected to the Commons as a member for Derby.[41] When some in the Commons proved uncooperative the king dissolved parliament on 5 May – four days before Hobbes completed the *Elements of Law*.[42] Not long afterwards, the Scots defeated Charles and forced him to summon the English parliament once more. This new assembly – which became known as the Long Parliament since it sat for many years – soon began to dismantle the king's regime, arresting some of his leading ministers and embarking upon a programme of institutional reform. Archbishop Laud, Heylin, Beale, and Maynwaring were amongst those called in question for what they had said and done during the 1630s. In view of such events, Hobbes decided to flee the country, returning once more to Paris (Aubrey 1:334). In April 1641 he wrote to Lord Scudamore – a close friend of Laud's and a former English ambassador to France – explaining why he had fled. 'The reason I came away', he said, 'was that I saw words that tended to advance the prerogative of kings began to be examined in Parlement'. He added that 'I know some that had a good will to have had me troubled'.⁴³

3: 1640–51

Between 1640 and 1642 political divisions increased within and outside parliament. One of the most contentious issues was church government. The policies which Archbishop Laud had pursued in the 1630s were widely resented, and some people now wanted to abolish episcopacy altogether. Others believed that bishops should be retained but that their powers should be greatly reduced. Many of those who shared Hobbes' broad political position argued that episcopacy had been instituted by God and that bishops held their powers by divine right, though they could not exercise them except with the consent of the monarch. In the summer of 1641 Hobbes wrote to the Earl of Devonshire, discussing the question. His letter reveals that he was keeping in close touch with English political developments. It also shows that he had no particular love for the bishops, referring to their 'Covetousnesse and supercilious behaviour', and to the 'abundance of abuses' which they and their supporters had recently committed. He was perfectly willing to countenance schemes which would place the government of the church largely in the hands of lay commissioners and he made it clear that he wanted the church wholly subordinated to the state. The dispute between 'the spirituall and civill power', he said, 'has of late more than any other thing in the world, bene the cause of civill warres in all places of Christendome'.⁴⁴

A few months after writing this letter, Hobbes completed the second version of his political philosophy, entitled *Elementorum philosophiae sectio tertia De Cive* (the third section of the elements of philosophy, concerning the citizen), and now usually referred to as *De Cive*. The dedication (to the Earl of Devonshire) is dated 1 November 1641. The book was written in Latin, for a Continental as well as an English audience. The style was simple and straightforward, resembling that of Bellarmine or De Dominis rather than the polished humanist Latin of Andrewes or Casaubon. *De Cive* was published at Paris, without the author's name (but with the initials 'T. H.' at the end of the dedication), in April 1642. Only a small number of copies was issued, but Mersenne was soon busy circulating the book amongst his learned friends and eliciting comments. Hobbes was to incorporate replies to criticisms made of the work in a second and much larger edition, published by the great Dutch firm of Elzevir early in 1647.

A great deal of *De Cive* was taken wholesale from the *Elements of Law*. The main differences between the two books were that *De Cive* omitted the discussion of human nature which took up the first thirteen chapters of the earlier work, and that it included a greatly expanded section on religion. The basic argument of this section was that the clergy are wholly subject to the state – the theme on which he had recently written to the Earl of Devonshire. The reason why Hobbes excluded the material on human nature from *De Cive* was that he now planned to deal with it in a separate volume. He intended to compose three works covering the whole field of philosophy. The first was to deal with matter and its general properties, the second with human nature, and the third with political philosophy: this was why *De Cive* was called the *third* section of the elements of philosophy. He brought *De Cive* out earliest because of its relevance to the debates which heralded the outbreak of civil war in England (DC 'praefatio ad lectores', 18).

At Paris during the 1640s Hobbes participated fully in the intellectual life of Mersenne's circle. In 1641 he conducted a debate by letter with Descartes (who was in the Netherlands) on problems of philosophy and optics (CM10:427–31, 488–99, 569–75, 589–90). Mersenne had asked Hobbes and others to comment on Descartes' recently completed *Meditations on first philosophy* (or metaphysics). The *Meditations*, along with the objections and Descartes' replies to them, were published in 1641. Hobbes did not have a very high opinion of Descartes, who in turn thought little of him.[45] Late in 1642 and early in 1643 Hobbes wrote a long reply to Thomas White's *De Mundo* (1642), discussing a variety of scientific and philosophical problems. This book remained in manuscript until 1973.

White was an unorthodox Catholic priest and a good friend of Sir Kenelm Digby. Digby had been imprisoned by the parliamentarians, but in 1643 was released and went to Paris. In the previous year the third Earl of Devonshire had likewise left England. From 1644 the Civil War began to go badly for the king, and many leading royalists sought refuge abroad. The war ended with defeat for the king in 1646. It was renewed in 1648, and again the royalists were beaten. Late in the same year, Parliament was purged of those likely to oppose the trial and execution of the king – an event known as Pride's Purge. The remaining members came to be called the Rump

Parliament. In 1649 Charles I was executed and for the next few years the Rump ruled in England. With Scottish help the dead king's son Charles II attempted to assert his claim to the throne in 1650, but he was finally defeated by Oliver Cromwell in September 1651.

After the important parliamentarian victory at Marston Moor in July 1644, the Marquis (as he now was) of Newcastle left England, accompanied by Sir Charles Cavendish, Bishop John Bramhall, and others. In April 1645 they arrived at Paris and resumed their contacts with Hobbes.[46] During the summer of 1646 Charles Prince of Wales – the future Charles II – reached the French capital. It was probably through the good offices of the leading royalist Henry Jermyn that Hobbes was hired to teach mathematics to the Prince – who later referred to him as 'the oddest fellow he ever met with'.[47] Hobbes also acted as tutor to other notables. These included the son and nephew of the wealthy poet Edmund Waller (who was related to John Hampden and by marriage to Oliver Cromwell).[48] Another and apparently rather inattentive student was the young Duke of Buckingham (who had been Aglionby's pupil). According to Aubrey, during one lesson Hobbes observed 'that his grace was at mastrupation (his hand in his codpiece)'.[49]

Teaching was not the only thing that kept Hobbes busy during these years. Mersenne published two collections of papers on scientific and mathematical subjects in 1644. Both contained contributions by Hobbes.[50] In late 1645 and early 1646 he wrote an English work on optics at the behest of the Marquis of Newcastle. A fair copy was made by William Petty[51] who remained a close friend of Hobbes henceforth, and who ranked him (along with Molière, Suarez, Galileo, Sir Thomas More, Bacon, Donne and Descartes) as one of the greatest systematic thinkers of recent times.[52]

At about the same time, Hobbes became involved in a controversy over determinism and free-will. He held that the universe consists of matter in motion, and that motion has been inexorably determined. Since everything that happens is caused, he argued, people do not possess free will. His opponent was John Bramhall, Bishop of Derry. The debate began at Paris in the presence of Newcastle, to whom Bramhall later sent a written version of his own opinion (which was that people do have free will). Hobbes wrote a response to Bramhall,[53] which was published – without Hobbes' permission – in 1654 (EW4:229–78). Bramhall's reply was printed in the following year, and in 1656 Hobbes responded with a point-by-point answer to the Bishop's book (EW5). A lengthy response by Bramhall to this latest volume was published with the date 1657, and with a long appendix dated 1658. The appendix attacked *Leviathan*, and accused Hobbes of atheism and impiety – charges to which Hobbes replied some ten years later (EW4:279–384).[54]

After the appearance of *De Cive*, Hobbes' friends were keen for him to publish the remaining two sections of his philosophy. Sorbière encouraged him to complete the project, and arranged for the publication of the 1647 edition of *De Cive* partly in the hope that it would lead Hobbes to finish the rest of his work.[55] What impressed friends like Du Verdus and Petty about Hobbes' political system was less the conclusions which he drew from it than the geometrical method it utilised.[56] In letters to Sorbière, Gassendi recorded his dissent from the religious views expressed in *De Cive*, and Mersenne, though praising the book, immediately went on to request the publication of the rest of Hobbes' philosophy.[57] Sorbière held that Hobbes' non-political works would be 'more pleasant' than *De Cive*, and that in those long-expected writings he would 'be able to make a slip or stray with less risk'.[58] Sir Charles Cavendish eagerly awaited the completion of Hobbes' trilogy, reporting on its progress to Pell.[59] In England, too, Hobbes' works on body and on human nature were avidly expected. Early in 1650 an unauthorised edition of the first part of the *Elements of Law* was printed for the Oxford bookseller Francis Bowman under the misleading title *Humane Nature*, with the intention of deceiving people into thinking that here at last was the second section of Hobbes' system.[60] In the event, the authentic second section (*De Homine* – Of Man) came out only in 1658, while the first section (*De Corpore* – Of Body) appeared in 1655.

One reason for the delay in completing the project was the drain on Hobbes' time caused by his commitments at Paris. As early as 1644 he told Sir Charles Cavendish that he was putting his philosophy 'in order'. Cavendish rightly feared that this might 'take a long time'.[61] In June 1646 Hobbes reported to Sorbière that he hoped to finish the trilogy within a year, and added that to gain more leisure he intended to retire into the country with his friend the nobleman Thomas de Martel.[62] This plan did not materialise, because shortly afterwards Hobbes was appointed to teach mathematics to Prince Charles – though according to Cavendish he still had leisure to pursue his philosophy.[63] In the summer of 1647 a serious illness left Hobbes unable to work for many weeks, but in November he wrote to Sorbière claiming that *De Corpore* would soon be finished.[64]

Eight months later Cavendish reported that Hobbes 'hath now leisure to study' (presumably since the Prince of Wales had left France for the Netherlands). He expressed the hope that the work would be finished within a year.[65] But in March 1650 he was still eagerly awaiting it. Shortly before that date the poet Sir William Davenant had sent Newcastle the preface to his projected poem *Gondibert*, along with a reply by Hobbes commenting on literature (EW4:441-58). 'I had rather read his philosophy', said Cavendish, 'which I hope he will ere long publish'.[66]

4: 1651-79

By 1650, however, Hobbes had for the moment shelved his project for a Latin philosophical trilogy and was hard at work writing a book in English on politics – *Leviathan*. By May he had completed thirty-seven chapters. Earlier that year he discussed the powers of bishops in correspondence with Robert Payne – which suggests that he was writing the third part of *Leviathan*. Hobbes held that bishops possessed no authority except as the sovereign's delegates. His position on this question had not changed much since *De Cive*, though *Leviathan* expressed his opinions more clearly and vigorously. Payne patently regarded Hobbes as a royalist and tried to persuade him to defend episcopacy – the form of church government which most royalists favoured. Hobbes refused. In Payne's mind the reasons for this were plain. Writing on 7 March 1650 he recorded that a letter from Hobbes to the Earl of Devonshire had recently been intercepted. In it Hobbes stated 'that he had lost the reward of his labours with the Prince by the sinister suggestion of some of the clergy'. Payne was 'very sorry to hear' that any clergyman had had 'the ill fortune to provoke so great a wit against the church'[67]

Easily the most obvious difference between *Leviathan* and *De Cive* is that the proportion of material devoted to religion and church-state relations is far greater in the former than the latter. One reason for this was that throughout the 1640s the question of the clergy's power was very hotly debated, and very divisive. Presbyterian ministers attempted to establish a rigid and intolerant ecclesiastical system in England, in which the opinions of the laity were to be vetted by churchmen. A major purpose of *Leviathan* was to refute Presbyterian ideas – and all theories which gave the clergy power over the state.

If Payne was right, then Hobbes had a personal motive for mounting the vigorous attack on clericalist ideas which takes up much of the last half of *Leviathan*. There were other reasons why the years around 1650 were a good time to publish books like *Leviathan*, which contained heterodox opinions on many subjects. Censorship was lax in England in these years, and Cromwell's victories had ensured the defeat of attempts to enforce rigid Presbyterianism. Unusual ideas could be expressed with relative impunity in the early 1650s, and anti-clerical opinions were especially fashionable. The clergy – or the ministry – were under attack from a variety of quarters, ranging from radical sectarians to that most learned of lawyers, John Selden. Bemoaning the sentence of death for treason passed against the 'godly Presbyterian minister' Christopher Love in 1651, Ralph Josselin – himself a minister – confided to his diary that 'the lord shatters us, and lays us low'. He frequently reverted to this theme, for example remarking in 1653 that it was generally said that 'now down go the ministers'. The minister George Lawson thought that *Leviathan*'s anti-clerical arguments were particularly appropriate 'in these Lunatick times' when 'simple giddy fools' were likely to believe Hobbes' doctrines though they were 'directly contrary to express Scripture'.[68]

Leviathan was published at London by May 1651.[69] *De Cive* had already received a hostile response from some royalists – including Bramhall (EW5:26) – but *Leviathan* aroused much greater antagonism, principally because of its claims on religion and church-state relations. Less than a year after its publication, Hobbes had returned to England and submitted to the republican government.[70] Before leaving France he had been forbidden to appear at the court of Charles II. He later recorded that some English clerics had told Charles that *Leviathan* contained principles which were heretical and which ran counter to the royalist cause. Once the king's protection had been withdrawn, he continued, he was compelled to return to England through fear of the Catholic clergy in France (Latin prose life, in LW1:xvii). Elsewhere he wrote that the recent assassinations by royalists of the parliamentarians Ascham and Dorislaus led him to fear for his own life and therefore to return to England, though he was unsure that he would be safe there (Latin verse life, LW1: xciii). Sir Edward Nicholas, a very well informed source, spoke of Hobbes' exclusion from court as a recent event in January 1652. He was happy that the 'father of atheists' had at last been removed from court since he had already done much to spread his atheism there.

Writing to Hyde, Nicholas mentioned the report that it was English Catholics, and especially Wat Montagu, who were responsible for the fall from royal favour of 'that great atheist'. Hyde, however, denied that the Catholics had been involved and took upon himself much of the credit 'for the discountenancing of my old friend, Mr H[obbes]'.[71]

The accusation of atheism was based upon Hobbes' distinctly odd views about God, and upon his denial that there are any incorporeal substances (since most people thought that God is an incorporeal substance). The charge of disloyalty was grounded on Hobbes' doctrine that members of a vanquished population could consent to the rule of their conquerors and thereby acquire an obligation to obey them. In 1650 the republican government in England had required adult males to take an Engagement, promising obedience to the new regime. The Engagement led to a pamphlet war. Hobbes contributed to this debate by drawing on ideas which he had already expressed in his earlier political writings to elicit the conclusion that people may obey a successful conqueror.

In some respects his argument resembled that of publicists for the republic – including Anthony Ascham (who was assassinated in 1650). Ascham had indeed been aware of Hobbes' theory as it was formulated in his earlier writings, and he used it to support his own position. But he also fully appreciated the differences between his views and those of Hobbes.[72] Hobbes was a thoroughgoing absolutist, while Ascham argued for limited and accountable governments. Writing against the important royalist Henry Hammond in 1650, another apologist for the new regime (who called himself 'Eutactus Philodemius') mentioned that Hobbes was 'a man of dangerous and unsound principles' on most questions, and also that he was one of Hammond's party – in other words, a royalist – but nevertheless cited him to confirm a point.[73] Hobbes' argument on conquest pleased adherents of the republic and annoyed Charles II's supporters, but it was a relatively minor part of his theory, and represented no sharp break from views voiced by pre-Civil War proponents of the divine right of kings.[74]

In England, Hobbes gained no rewards from the republican government but was also largely unmolested by it or by the succeeding regimes of Oliver and Richard Cromwell. *Leviathan* was, indeed, reported as an atheistical work to a committee of parliament in 1657, but no action seems to have followed.[75] He resumed his connection with the Earl of Devonshire but spent much of his time in London, where his friends included Selden and William Harvey.[76] Though Hobbes suggested in *Leviathan* that the best form of church government was Independency, there is little reason to think that he stood particularly high in the good graces of the Independent leadership during the 1650s. John Owen, perhaps the most important theorist of Independency in those years, did indeed admit that there were some good things in *Leviathan*. But he strongly objected to Hobbes' deification of the magistrate, and claimed that he 'spoiled all' in the fourth part of the book, which inveighed against the clergy and the universities. Viscount Saye and Sele – who had been a leading lay Independent in the 1640s though he took little part in politics during the 1650s – equated 'hobbs and athiests', and contrasted them with 'good men and christians'.[77]

At a time when the ministry was under threat from radicals – soon including the Quakers – it was natural enough that churchmen should resent the attacks on the clergy mounted by intellectuals like Hobbes and Selden. In a letter of 1651 which discussed Hobbes' *De Cive* and Selden's *De synedriis*, Brian Duppa (Bishop of Salisbury) bemoaned the fact that such noted scholars were now attacking the church. '[T]he Church is likely to be stript by learned hands', he commented, 'which seemes sadder to mee than all her sufferings from the rabble'. Duppa nevertheless admired aspects of Hobbes' work, claiming that the attempt to create a geometrical science of politics was 'the highest reach of man's witt'. He found many things in *Leviathan* 'said so well that I could embrace him for it', but was highly irritated by Hobbes' jesting approach to religion and doubted that 'he was ever Christian'. Thomas Barlow – librarian of the Bodleian at Oxford, and later Bishop of Lincoln – also liked much about *Leviathan*, though he hinted that he thought some of it was heretical.[78]

After the Restoration of Charles II in 1660, the bishops were reluctant to forgive Hobbes for his attacks on the church and for his heterodox religious views. Others, such as Hyde, were anxious to emphasise the moderate face of monarchy, and to demonstrate that royalists in general dissented from Hobbes' absolutist opinions by attacking them. Hobbes' reputation for heresy and atheism made it prudent for those whose views were (on some questions) uncomfortably close to his to dissociate themselves from him, and it is likely that this is why he was never elected to the Royal Society.[79] In 1666 a committee of the House of Commons was authorised to investigate charges of atheism levelled against *Leviathan*, and

Aubrey records a report that 'some of the bishops made a motion to have the good old gentleman burnt for a heretic' (1:339).[80]

Hobbes had powerful friends – including the king himself, and the Earl of Arlington – who saw to it that he came to no harm. But he was forbidden to publish works on sensitive political questions in England, and it was at Amsterdam that the revised Latin edition of *Leviathan* appeared in 1668 – an edition that toned down or omitted some of the more controversial passages in the original English version. Hobbes responded to the charges of heresy by writing *An historical narration concerning heresy and the punishment thereof*, in which he argued that statute law did not permit the execution of heretics in England (EW4:385–408).[81] The same theme cropped up again in his *Dialogue between a philosopher and a student of the common laws of England*, which asserted the superiority of statute to common law, and developed some of the criticisms which he had already levelled in *Leviathan* against the ideas of Sir Edward Coke – an extremely influential legal writer and a major opponent of absolutism under James I and his son.

These works were not published until after Hobbes' death in 1679, and the same goes for two other books discussing topics he had already broached in *Leviathan*. One was a Latin verse history – the *Historia Ecclesiastica* (LW5:341–408) – dealing with the evils wreaked by ambitious clerics upon religion and the state from the earliest times until the sixteenth century. The second was *Behemoth or the Long Parliament* – a history of England from before the Civil War until the Restoration. The basic attitudes that informed these works were largely unchanged from Hobbes' earlier writings. In other respects, too, he kept up his former interests in old age. During the 1650s he entered into controversy on mathematics with John Wallis, and the debate – which harmed his scientific reputation – was still alive in the 1670s.[82] At the start of his literary career he published a translation of Thucydides. By 1674 he had completed a translation of the *Odyssey* and *Iliad* of Homer into English verse. It was probably in 1679 that he wrote a brief manuscript which applied his political theories to the problems raised by the Exclusion Crisis, arguing that a king may but is not obliged to exclude an incapable heir from the succession.[83] In December 1679 he died, at the age of ninety-one.

2
The Law of Nature and the Natural Condition of Mankind

Hobbes held that the universe consists of nothing but matter in motion – or at rest. He claimed that human psychology may be reduced to physical laws. The opening chapters of *Leviathan* are devoted to illustrating and confirming this thesis. He also held that people can understand human nature by introspection, arguing (as Montaigne and Descartes had argued) that by examining our own thoughts and passions we may discover truths about the thoughts and passions of everyone (Lev Introduction 2). His political doctrines were based upon his theory of human nature – a theory which *De Corpore* (on body, or matter) and *De Homine* (on man) were intended to vindicate, but which (in Hobbes' opinion) was sufficiently confirmed by self-examination and by everyday experience. It was because his political theory was grounded in principles evident from experience that Hobbes felt able to publish *De Cive* before the other two parts of his philosophical trilogy (DC preface to the reader, 10, 19).

'[O]f the voluntary acts of every man', said Hobbes, 'the object is some *Good to himselfe*' (Lev 14: 93/66). He is often described as a psychological egoist, and he certainly delighted in expressing cynical opinions about human nature. But he did not exclude the possibility of altruism, listing benevolence, good will and charity amongst the passions (Lev 6: 41/26). True, some passages in his writings suggest that he thought people acted benevolently only from self-regarding motives; but the text is ambiguous, and it is unclear that his teaching differs much from that of earlier thinkers – including Aristotle – on this point.[1]

Hobbes argued that '*Good* and *Evill*, are names that signifie our

Appetites, and Aversions' (Lev 15: 110/79). Virtually everyone thought that people desire what seems good to them. Most went on to argue that a number of things are in fact good and that reason tells us what they are. Hobbes claimed that reason guides us on how to achieve what we desire, but says nothing about what we ought to desire. The objects of individuals' desires vary in accordance with their personal characteristics. But all, at least ordinarily, desire self-preservation. Since peace enhances our prospects of preserving ourselves 'all men agree on this, that Peace is Good' (Lev 15: 111/80). So an objective science of politics may be constructed by examining what is necessary for people to attain peace.

The natural condition of mankind outside civil society, said Hobbes, is one of misery (Lev 17: 117/85). By nature, he claimed, people are equals, at least in the sense that even 'the weakest has strength enough to kill the strongest, either by secret machination, or by confederacy with others' (Lev 13: 87/60; cf. DC1:3; EL1:14:2). Approximate equality makes it rational for everyone to fear everyone else in the state of nature, and fear gives rise to war. So outside the civil state people 'are in that condition which is called Warre; and such a warre, as is of every man, against every man' (Lev 13: 88/62; cf. DC1:12; EL1:14:11). In these circumstances, the pursuit of cooperative enterprises is impossible and people lack all the amenities of civilized living. Their lives are 'solitary, poore, nasty, brutish, and short' (Lev 13: 89/62; cf. DC1:13; EL1:14:12).

In the state of nature, Hobbes tells us, each individual has the 'Right of Nature' or '*Jus Naturale*', which is 'the Liberty . . . to use his own power, as he will himself, for the preservation of his own Nature; that is to say, of his own Life; and consequently, of doing any thing, which in his own Judgement, and Reason, hee shall conceive to be the aptest means thereunto' (Lev 14: 91/64; cf. DC1:7; EL1:14:6). By nature, we shun what harms us, and especially death, 'the greatest of natural evils' (DC1:7: 'maximi malorum naturalium, quae est mors'; cf. EL1:14:6).[2] But if every individual exercises the right of nature and does literally whatever he happens to think will promote his own preservation, the result will be a state of warfare. Warfare, however, does not help to preserve us.

So exercising the right of nature is no sufficient means of securing self-preservation. Hobbes held that reason set down certain rules 'by which a man is forbidden to do, that, which is destructive of his life, or taketh away the means of preserving the same; and to omit, that, by which he thinketh it may be best preserved' (Lev 14: 91/64;

cf. DC2:1). He termed these rules the laws of nature. By following these laws, he claimed, men could leave the state of nature and erect a commonwealth which would allow them to live in security.

According to Hobbes, the 'Lawes of Nature are Immutable and Eternall' (Lev 15: 110/79). '[W]hat they forbid', he said, 'can never be lawful; what they command can never be unlawful' (DC3:29: 'quod vetant nunquam licitum esse potest; quod iubent nunquam illicitum'). However, circumstances can arise in which it is not rational to act according to the laws of nature. If we see that other people are not putting the laws into practice, it might be foolish for us to do so. For example, a man who kept all his promises while everyone else broke theirs 'should but make himselfe a prey to others, and procure his own certain ruine'. This would be 'contrary to the ground of all Lawes of Nature, which tend to Natures preservation' (Lev 15: 110/79; cf DC3:27). In the state of nature it would be self-defeating to keep contracts if we suspected – or had a 'just fear' (*justus metus*: DC2:11, 13:7) – that those with whom we had contracted were going to break their part of the bargain. The laws of nature are immutable and eternal, then, not in the sense that we are bound always 'to the putting them in act' but in the sense that 'they bind to a desire they should take place' (Lev 15: 110/79). We should always stand in readiness to put them into practice (EL1:17:10; DC3:27), and should in fact do so whenever circumstances arise in which such a course would be safe. The laws of nature themselves guide us to an understanding of how to create such circumstances.

In each of his three major treatments of the subject (the *Elements of Law*, *De Cive* and *Leviathan*) Hobbes gave a slightly different list of the laws of nature. In the *Elements*, for example, there is no hint that drunkenness is against the law of nature, while in *De Cive* we learn that it is an infringement of the last of the twenty laws (DC3:25).[3] In *Leviathan* it has again dropped from the list (now reduced to nineteen), though it is briefly mentioned (along with 'all other parts of Intemperance') in a supplementary passage (Lev 15: 109/78). Near the end of *Leviathan*, Hobbes introduced a new law of nature which was of obvious relevance to questions raised by the English Civil War – the law that we must do all we can to protect in war the authority by which we are protected in peace (Lev 'Review and Conclusion', 484/390). But the principles set down as laws of nature are largely the same in all three works. Hobbes admitted that his lists of laws were incomplete, and argued that a summary of all the laws of nature was provided by the Golden Rule: *'Do not that to another which thou wouldest not have done to thy selfe'* (Lev 15: 109/78–9; cf. DC3:26, EL1:17:9).[4] The laws of nature, he said, were 'dictates of Reason', and 'Conclusions, or Theoremes concerning what conduceth to the conservation and defence' of people. In this sense they were not strictly laws at all since 'Law, properly is the word of him, that by right hath command over others'. However, he maintained that the laws of nature could also be found 'in the word of God, that by right commandeth all things'; so in this sense they were 'properly called Lawes' (Lev 15: 111/80). In *De Cive* he devoted a chapter to showing that Scripture confirmed natural law, which was therefore divine (DC4; cf. EL1:18).

The fundamental law, from which the others are deductions, is that we should *'endeavour Peace'*, but when we cannot obtain it *'seek, and use, all helps, and advantages of Warre'* (Lev 14: 92/64; cf. DC2:2). Since peace is unattainable as long as people exercise the right of nature – the right to do absolutely anything that they think conduces to their preservation – it is necessary that they lay down or transfer this right. But no one has any reason to part with rights unless others do so too. What is needed to escape from the state of nature, therefore, is a mutual transference of rights by which each person agrees to hand over his rights to a particular individual or assembly on condition that the rest do so too. Hobbes defines a contract (*contractus*) as 'the mutuall transferring of Right', and says that a pact or covenant (*pactum*) is a contract in which one or both parties undertake to perform their part of the bargain not immediately but in the future (Lev 14: 94/66; cf. DC2:9; EL1:15:8–9). It is a law of nature, he says, that *'men performe their Covenants made'* (Lev 15: 100/71), and it is by obeying this law that they are able to institute a commonwealth.

On Hobbes's account, fear is a major motive impelling people in the state of nature to make the covenant by which a sovereign is set up. In arguing like this, Hobbes commits himself to the position that 'Covenants entred into by fear, in the condition of meer Nature, are obligatory' (Lev 14: 97/69). Ordinarily, he claims, we are bound to fulfil agreements which we entered into under threat. For instance, if I am captured by a robber and he agrees to spare my life on condition that I promise to pay him a ransom at some future date, I am obliged to keep my promise (Lev 14: 97/69). Similarly, if I covenant to obey a conqueror as my sovereign provided that he spares my life, I am obliged to do so (Lev 20: 141/104).

So sovereigns may arise not only by institution but also by conquest. In Hobbes's view it is not the fact of conquest but the subsequent agreement between conquered and conqueror which turns their relationship into that of subject and sovereign. He also argued that in the natural condition of mankind fathers may and generally do hold sovereign power over their children. He analysed this relationship, too, in terms of consent. The father's sovereign power over the child, he said, is derived 'from the Childs Consent, either expresse, or by other sufficient arguments declared' (Lev 20: 139/102). So commonwealths, and sovereigns, can arise either by institution, conquest, or from the relationship between father and child. But in each case it is consent which institutes sovereignty and subjection.

In Hobbes's theory, the reason why people consent to subject themselves to a sovereign is that by so doing they will obtain security, since the sovereign will protect them from attack by their fellows. 'The end of Obedience', he said, 'is Protection' (Lev 21: 153/114). Two consequences flowed from this. Firstly, if the sovereign lost his power to protect us – say, because he was defeated by a foreign enemy or by domestic rebels – then we would be absolved from our obligation to obey him, and would resume the right of nature to do all that we judged necessary to protect ourselves. Secondly, if someone other than the sovereign – say the rebel leader – was in fact protecting us, it would be rational for us to acknowledge him as our sovereign (Lev 21: 153–4/114; 'Review and Conclusion' 484/390; cf. DC7:18; EL2:2:15). In *Leviathan*, Hobbes placed great stress upon the link between protection and obedience, and spelled out some of the practical implications of the doctrine for contemporary England – in which the Rump had come to replace the king as the people's protector.

Hobbes castigated contemporary moral theorists for failing to treat their subject as systematically as geometers did theirs. Moral philosophy, he claimed, had not advanced since ancient times, and its practitioners failed to reason accurately from first principles, instead disguising their rashly received opinions in fine language (DC epistle dedicatory, 6–8). His own works were intended to remedy this state of affairs by providing a rigorously logical system in which the conclusions followed inexorably from first principles, and especially from the principle of self-preservation. As we have seen, Hobbes claimed that his own system was not only correct but also highly original, asserting that moral science was no older than his *De Cive* (EW1:ix; cf. LW1:cv). Even if we take this claim at face value, however, it by no means follows that Hobbes' views were unrelated to those of his contemporaries.

In fact, as we shall see, a large proportion of his ideas were similar to or identical with those of earlier writers. Self-preservation, the pre-civil state of mankind, the law of nature, contract and consent, conquest, fatherhood, and the link between protection and obedience all feature prominently in the political works of Hobbes' contemporaries. Though his theory purported to be a series of deductions from first principles, many of his doctrines were related in only the vaguest way to his premises, but were linked far more closely to the exigencies of contemporary debate. It makes sense to begin our discussion with the central principle of his system – self-preservation.

1: SELF-PRESERVATION

Self-preservation is the cornerstone of Hobbes' theory. The principle, he argued, underlay both the right of nature and its laws. The laws of nature, said Hobbes, are 'derived from a single dictate of reason, exhorting us to take care for our preservation and safety' (DC3:26: 'ab unico rationis, nos ad nostri conservationem & incolumitatem hortantis, dictamine derivata'). The principle of self-preservation played a major part in the thinking of Hobbes' predecessors. All agreed in giving it great moral weight. But no one attempted to ground a whole ethical and political system upon it in so rigorous a fashion as Hobbes. Much of the peculiarity of Hobbes' political theory arises from the emphasis that he placed on self-preservation. To give one striking instance, he was a royalist who accepted and indeed welcomed one of the key doctrines of anti-royalists – that people could defend themselves even against the king himself.

A commonplace idea in medieval and early modern Europe was that the life of a man is not his own to do with as he sees fit. No individual, said Clarendon, 'is lord of his own life' (*dominus vitae suae*), taking precisely the same stance as Suarez, who asserted that it is God who is 'the lord of life' (*dominus vitae*).[5] It was the Lord who had given us life, and it was He who disposed of our lives; so no one was entitled to give up life unless the Lord willed it. Self-preservation, then, was a duty towards God unless He willed otherwise. His will was sometimes expressed directly, in that He could allow us to die by mortal illness or accident. It could also be expressed through the institution of laws punishing criminals with death. If a criminal is justly sentenced to death, so the common view ran, then he is obliged meekly to accept his sentence. Hobbes was virtually alone in rejecting this idea. He claimed that criminals could resist all the way to the gallows (Lev 21: 151/111–12), for no one could forfeit the liberty of preserving himself.

The extremely influential medieval theorist St Thomas Aquinas claimed that moral truths are ranged in an order which corresponds to mankind's natural inclinations, and that the most basic principles of morality are connected with the preservation of life. All substances, he argued, seek their own preservation, and this is true of man among the rest. But he did not attempt to base the whole of his moral system upon this principle. Rather, he distinguished between three aspects of human nature, and argued that each gives rise to moral truths. Firstly, human individuals (like all other beings) seek their own preservation. Secondly, however, men (like other animals) seek the preservation of the whole species as distinct from individual members. Thirdly, man possesses a rational and spiritual nature which he does not share with other animals, and which teaches him truths about God and society.[6] The principles generated by human nature in the second and third of these three aspects could cut against the rule of self-preservation.

Aquinas did indeed argue that people are bound to take care for their own preservation before that of anyone else: 'a man has more obligation to provide for his own life than for that of another'. Though killing is ordinarily wrong, he claimed, it is permissible to defend yourself by force against unjust attack even if the consequence is the death of the attacker. He approvingly quoted the Roman Law maxim that 'it is licit to repel force with force, provided that you moderate the force to what is necessary for your defence'. The 'right of self-defence is the greatest of rights', said Suarez. Indeed, it was generally held that in most circumstances self-defence is not just permissible but obligatory. 'By the law of nature', said the influential sixteenth-century thinker Jacques Almain, 'everyone is bound to preserve himself'.[7]

Such ideas were commonplace in England during the Civil War. Almain's discussion was cited by the cleric William Bridge in a pamphlet published in 1643 to justify parliament's resistance to the king. According to Bridge, 'It is the most natural work in the world for everything to preserve itself'. 'The law of nature and common prudence', declared a pamphlet of 1647, 'enjoins every man to preserve himself'. 'Self-preservation is by the law of nature', said the lawyer John Cook in the same year, and he added that 'no man can give away the right of defending his own life until he hath forfeited it'. The conventional view was that people could forfeit the right of self-defence by committing crimes warranting the death penalty, but that normally they had a duty to defend themselves against attack. If they failed to do so, they would be guilty of their own destruction. The ban on murder was seen as ruling out suicide, and this in turn entailed a duty to prevent your own death. Parliamentarians argued that this duty applied to communities as well as individuals. 'The people can no more resign power of self-defence, which nature hath given them, than they can be guilty of self-murder', asserted the leading Scottish Presbyterian Samuel Rutherford.[8] Suarez had said much the same thing, arguing that if violent defence against the king is licit to save your own life, it must be much more so to preserve the community.[9]

This position was a staple item in parliamentary propaganda during the Civil War. Every 'private man may defend himselfe by force', said Henry Parker in his famous *Observations* of 1642, 'if assaulted, though by the force of a Magistrate or his owne father'. He claimed that this was according to 'the clearest beames of humaine reason, and the strongest inclinations of nature'. All individuals, asserted William Prynne in the same year, may licitly defend themselves against unlawful attack even by a husband or king. The idea was used to support the claim that the whole community could likewise defend itself, and therefore that parliament – which represented the community in England – was justified in fighting the royalists. There was, as we have seen, nothing new about the notion. Almain, on whose work Bridge drew, used the analogy between individual and public rights to justify resistance by a community against a tyrannical ruler.[10]

In the Civil War, royalists rebutted the argument by rejecting the idea that individuals have a right of self-defence against the king, and by claiming that the notion was a recipe for anarchy. They also denied the community any analogous right of self-defence. According to Michael Hudson (a chaplain to Charles I), our duty to obey the civil magistrate is of a higher order than the obligation to defend ourselves. So we may never defend ourselves against the king. In 1643 Henry Ferne argued that to allow subjects a right of self-defence against their rulers would be to provide a remedy 'worse than the disease, and more subversive of a state than if they were left without it'. A private individual, he affirmed, did indeed have the 'power of selfe preservation against the force of another private person'; but 'this power' was 'yeelded up in regard of the Civill power . . . and not to be used against persons indued with such power'. Individuals benefited from the rule of a king, who protected them against attack by others. But the king's ability to protect his subjects would be undermined if individuals were empowered to resist him whenever they thought they were being unjustly attacked. So individuals were obliged *not* to defend themselves against the king.[11]

The important royalist propagandists Henry Hammond and Dudley Digges argued in much the same way.[12] So too did pre-Civil War writers. In about 1616 Hugo Grotius claimed that the maxim *vim vi repellere licet* applies only between equals, and not against a superior (such as the king): 'violent defence though lawful against an equal is unlawful against a superior'.[13] In a work published in 1622 the cleric David Owen similarly stated that a private individual may not defend himself against a magistrate, even in a case of necessity, taking issue on this question with the influential resistance theorist David Pareus. Nine years earlier, Thomas Preston – defending James I's oath of allegiance against papalist attack – admitted that one individual could use force to resist the assault of another, but denied that such resistance was ever lawful against a magistrate. There was, he said, no 'licence of defending oneself by arms' in this case 'lest greater evils result for the community, and the common peace of the citizens be unjustly destroyed by tumults and seditions'.[14]

In the late sixteenth century some theorists who took a generally high view of the power of kings and who rejected notions of popular sovereignty, nevertheless allowed resistance to the monarch in extreme circumstances. William Barclay in his *De regno* of 1600 famously admitted that the commonwealth could defend itself by force against direct attack by the prince.[15] Locke was only one among many later writers to remind the world that this 'great Assertor of the Power and Sacredness of Kings' granted that 'Self-defence is a part of the Law of Nature' and that it could not 'be denied the Community, even against the King himself'.[16] But later absolutists avoided Barclay's concessions and asserted a duty of total non-resistance to kings. Hobbes parted from their views in drastic fashion. His critics pounced on the consequences of his claim that individuals are entitled to defend themselves by force against the king. This liberty of self-defence, said Clarendon, was 'utterly inconsistent with the security of prince and people'. The doctrine, declared Bramhall, opened 'a large window . . . to sedition and rebellion'.[17]

Parliamentarians used the idea that people have a right (or duty) to defend themselves in order to justify resistance to kings in a wide range of circumstances. Royalists suspected that Hobbes' principles could be employed to the same purpose.[18] He grounded his whole system upon individual self-preservation, and so evidently could not adopt the conventional royalist argument that it made sense for individuals to lay aside the right to defend themselves against their monarch. In Hobbes' opinion, failure to defend oneself was a sign of insanity, not sense.[19] He denied that the right of self-defence led to anarchic consequences, however, for he insisted that it could be used only in a very limited set of circumstances, and he strongly contrasted it with the truly anarchic right of nature which people possessed wherever there were no governments.

2: THE STATE AND THE RIGHT OF NATURE[20]

There was nothing particularly novel about the term or the concept of the state of nature. The term was employed by the important Catholic thinker Jean Gerson in the fifteenth century, by the Jesuit Molina in the late sixteenth, and by Grotius in the early seventeenth.[21] The concept was implicit in the thinking of all those who believed that government arose from the consent of the governed. Not everyone did believe this, for as we shall see some thinkers held that states were essentially the same thing as families, and that within families the authority of parent over child did not arise from consent. Since families had existed from the time of Adam onwards, these writers argued, there had always been civil states. From this perspective it made little sense to talk about a state of nature – a state which had existed before the commonwealth or civil society, or *the* state. But only a relatively small number of thinkers equated the state with the family. For most theorists, then, it was perfectly sensible to ask what things had been like before people consented to live together under governments – or what they would be like if people had not in fact consented to live in this way. Many writers did pose just these questions.

The answers they gave make it clear that Hobbes was not alone in conceiving of the state of nature as an unpleasant place. Suarez considered what things would be like without political society. In such circumstances, he argued, people would lack the skills and knowledge necessary for civilized life, and families would be 'divided against one another' so that 'peace could scarcely be preserved amongst men'. There would, he said, be 'the greatest confusion' in a community which lacked government. So human nature itself made government necessary. In a sermon preached in 1621, the puritan cleric Robert Bolton vigorously stressed the horrors of life without government: 'murder, adulteries, incests, rapes, roberies, perjuries, witchcrafts, blasphemies' and a host of other evils 'would overflow all Countries. We should have a very hell upon earth, and the face of it covered with blood, as it was once with water'. 'Remove government', said the preacher Henry Symons, 'and nations will quickly become dens of beasts for prey, slaughter houses of blood'. Without government, said the great Elizabethan theorist Richard Hooker, 'strifes and troubles would be endlesse'. Such sentiments were utterly conventional. Hobbes' comments on the horrors of life in the state of nature are amongst the most famous passages in his works; they are also amongst the least original.[22]

Nor was Hobbes alone in distinguishing between the law and the right of nature. The distinction was drawn by the fifteenth-century humanist Lorenzo Valla,[23] and it featured largely in the theories of two of Hobbes' contemporaries, Dudley Digges and Jeremy Taylor. Hobbes asserted that 'they that speak of this subject, use to confound *Jus*, and *Lex, Right* and *Law*'. He added that 'they ought to be distinguished; because RIGHT, consisteth in liberty to do, or to forebeare; Whereas LAW, determineth, and bindeth to one of them: so that Law, and Right, differ as much, as Obligation, and Liberty; which in one and the same matter are inconsistent' (Lev 14: 91/64; cf. EL2:10:5; DC14:3). Dudley Digges expressed similar views in much the same terms, arguing that '*Right* and *Law* differ as much as Liberty and Bonds: *Jus*, or right not laying any obligation, but signifying, we may equally choose to doe or not to doe without fault, whereas *Lex* or law determines us either to a particular performance by way of command, or a particular abstinence by way of prohibition'.[24]

Taylor likewise argued that '*Jus naturae* [the right of nature] and *lex naturae* [the law of nature] are usually confounded by divines and lawyers', insisting that 'the right of nature, or *jus naturae* is no law' but rather 'a perfect and universal liberty to do whatsoever can secure me or please me'.[25] Both Taylor and Digges distinguished between *jus* and *lex* in much the same way as Hobbes. Both spoke of life under the *jus naturae* as a state of complete liberty. Taylor took much of his doctrine from John Selden's *De jure naturali*, first published in 1640. Selden also talked of a state of absolute liberty. The notion was well enough known for the leading parliamentarian John Pym to be able to assert in 1641 that if there is no law 'every man hath a like right to any thing'.[26]

Selden, Taylor and Digges encountered little criticism for their views on the right and state of nature. By contrast, Hobbes' doctrines on the same questions were vigorously attacked by his opponents. For despite the orthodox elements in his account, Hobbes added novel twists to the doctrines. These resulted principally from the special status he gave to self-preservation. In Hobbes' account, people were not naturally sociable but were motivated to enter civil society largely through fear for their lives. This flew in the face of one of the most widely accepted beliefs of his contemporaries. Moreover, Hobbes argued that before they entered society and set up governments, people were entitled to perform any act which they thought would further their preservation. This included a wide range of acts which were usually considered immoral. Others claimed that people joined together in commonwealths because they were naturally sociable, and that even before they came together in political societies they possessed moral intuitions which told them what was right and wrong. They did not, indeed, always act on these intuitions, for the Fall had corrupted human nature, and people now easily slipped into wicked ways. It was for this reason, so the usual line went, that coercive power was necessary to maintain order amongst them. But they did possess knowledge of moral truths and were morally obliged to act on this knowledge. Hobbes, by contrast, claimed that it was not natural sociability but calculations of how best to provide for self-preservation which led people to enter commonwealths, and that there were no reliable moral intuitions except those deducible from the principle of self-preservation. His critic Tenison (who later became Archbishop of Canterbury) detected the influence of Gassendi's revived Epicureanism on Hobbes' views at this point.[27]

Certainly those views were very different from orthodox thinking.

Hobbes attacked the conventional picture of natural sociability at the very beginning of *De Cive* (DC1:1–2). He rightly observed that most of his contemporaries followed Aristotle in holding that man is naturally sociable. They assumed that just as bees naturally live in ordered societies, so do men; the analogy between the beehive and the human commonwealth was indeed commonplace. Hobbes held that it was also in some respects profoundly misguided. Bees, he claimed, are uncompetitive creatures, amongst whom there is no distinction between the good of the individual and of the whole hive. Lacking reason, they find nothing to criticise in the administration of their societies, and lacking language – 'the trumpet of war and sedition' ('tuba . . . belli . . . & seditionis': DC5:5) – they are unable to persuade each other that they are misgoverned. None of these things, he said, was true of men (EL1:19:5; DC5:5; Lev 17: 119–20/86–7).

Hobbes was doing nothing original in attacking the idea that arguments by analogy can provide a secure basis for political knowledge. The notion that analogies are a very low-level form of political argument was frequently expressed.[28] But Hobbes' contention that people are not naturally sociable was certainly seen as novel by his contemporaries. While others believed that a number of principles of conduct – including sociability – are built in to human nature, Hobbes' case centred on the claim that all such precepts flow from and are subordinate to individual self-preservation. People institute commonwealths, he argued, not because sociability is intrinsic to their natures, but because they correctly calculate that self-preservation cannot be adequately safeguarded otherwise. Where there is no commonwealth each individual possesses the right of nature, 'that every man may preserve his own life and limbs, with all the power he hath' (EL1:14:6; cf. DC1:7–8; Lev 14: 91/64).

In explaining why the resulting state of affairs – the state of nature – was one of radical insecurity, Hobbes drew attention to a number of common or universal characteristics of men. He argued that some individuals foolishly overestimated their own powers, failing to note that men are approximately equal in ability. Led by vainglory, such people sought to dominate their fellows (EL1:14:3; DC1:4). Hobbes made it clear that vainglory was irrational and that the aggressive actions of vainglorious men in the state of nature were blameworthy. The activities of these belligerent people made it prudent for moderate men to stand on their guard and be willing to hurt others in order to defend themselves (DC1:4). Indeed, anticipation was the best form of defence (Lev 13: 87–8/61), so it would be rational for moderate men themselves to become aggressors in the state of nature.

But Hobbes did not think that the deeds of the vainglorious were the only cause of instability in the state of nature. Competition for limited resources would also bring conflict (EL1:14:5; DC1:6; Lev 13: 87/61). While the aggression of vainglorious men resulted from their *failure* to perceive the equality of people, those who did recognise such equality would themselves be led to act aggressively since they would see that aggression could lead to success: '[f]rom this equality of ability, ariseth equality of hope in the attaining of our Ends' (Lev 13: 87/61). Again, conflict would result from the fact that men derive pleasure from comparing themselves favourably with their fellows. This leads them to perform actions which make clear that they hold others in contempt (DC1:5). But 'every man looketh that his companion should value him, at the same rate he sets upon himselfe' and 'naturally endeavours' to extort esteem from others (Lev 13: 88/61).[29]

Hobbes' doctrine that by nature people are approximately equal was not particularly novel. Though Filmer rejected the thesis of natural equality he noted that it was widely accepted.[30] Again, Hobbes' position on the state of nature in some ways resembles the traditional notion that coercive power is made necessary by the Fall, which corrupted human nature and led people to follow their emotions rather than reason.[31] True, Hobbes had little to say about the Fall. But like more conventional thinkers he argued that coercion is necessary because people are prone to act irrationally. Reason, Hobbes says, demonstrates that it is expedient for us to agree together to set up an authority which will resolve conflicts between us. But passion inclines us to break our agreements unless they are enforced by coercive means.

Despite the traditional elements in his thinking, Hobbes was frequently attacked for his account of human nature. Descartes thought that the principles he expressed in *De Cive* were 'very wicked and very dangerous, in that he supposes that all men are evil, or that he gives them reason to be so'. According to Clarendon, Hobbes described human nature as worse than that of beasts, for even beasts did not prey upon their own kind as Hobbes supposed men would naturally do.[32] Calvinists took a pessimistic view of human nature. It might be suggested that Hobbes' opinions on this topic were influenced by Augustinian and especially Calvinist attitudes. Certainly, Calvin held that fallen human nature is so corrupt that without God's grace we can do nothing to merit salvation; but his political theory was not Hobbesian. '[M]an', he asserted, 'is by nature a social animal' who is 'disposed, from natural instinct, to cherish and preserve society'. The minds of all men, he argued, 'have impressions of civil order and honesty'. Though 'headlong passion' did indeed lead some to act unjustly, for example by stealing, even these people continued to be aware of the principles of justice.[33]

The idea that all men are at least sometimes aware of eternally binding moral principles other than self-preservation lay at the root of attacks on Hobbes' picture of the state of nature. According to Clarendon, people possess an 'instinct towards justice' which is independent of self-interest and which distinguishes us from 'the beasts of the wilderness'.[34] Tenison claimed that in the state of nature men would have to put into execution the 'natural laws of good and evil' which included many duties unrelated to self-preservation. Even if the state of nature was, as Hobbes supposed, a state of war, Grotius had shown that not everything was licit in war. According to Edward Stillingfleet (later Bishop of Worcester), 'in a state of absolute liberty' before the institution of governments, people would be bound by 'the obligatory laws of nature'. He drew on the work of John Selden to confirm his views.[35]

As we saw, Selden posited an original state of absolute liberty, where no laws operated. This sounds like Hobbes' state of nature. But there was one major difference. Selden stressed that this state of absolute liberty was purely hypothetical, and insisted that mankind is always bound to execute God's laws of nature. Indeed, Selden's point in positing an original state of complete freedom was to show that even there certain universal moral principles must apply. For how could free people leave the state of liberty and institute governments except by agreeing to do so? Such agreements would have no effect unless there existed from the start a binding moral rule that we must keep our agreements. So even if we posit an original state of absolute freedom we must admit the existence of at least one universally binding moral principle to explain how we can leave that state. Selden claimed that there are in fact a number of such rules and that they are independent of self-interest, inveighing against those who thought that all moral rules are reducible to individual interests, and vigorously asserting that there are things which are '*per se* just or unjust'. The 'state of absolute liberty', said Stillingfleet, is 'an imaginary state, for better understanding the nature and obligation of laws'. Selden himself argued that man, like other animals, is naturally sociable.[36]

Digges posited an original state of complete freedom. So did Taylor. The right of nature, they argued, was for all to take whatever they desire. 'Whatsoever we naturally desire', said Taylor, 'naturally we are permitted to'. But he insisted that this right of nature was an abstraction, describing what we could do if we were not subject to law. '*Jus naturae*', he said, is 'a negative right, that is, such a right as every man hath without a law'. In fact, he claimed, all people are subject to a law which restrains their original hypothetical liberty. This is the law of nature, which is 'a transcript of the wisdom and will of God written in the tables of our minds'.[37] In the theory of Selden, Taylor and the rest, the state of absolute liberty abstracts from all law, including the law of nature. It is an imaginary state, illustrating what things would be like if no laws at all applied. In Hobbes' theory, by contrast, the state of nature abstracts only from coercive authority. It is a state where there is no sovereign, but where the law of nature *does* apply. What Hobbes is saying is that where there is no government it is fine to kill people if doing so accords with your plans for self-preservation. In the opinion of the others, killing people is usually contrary to the law of nature and therefore wrong, though if there were no law of nature we would have a right to kill. This explains why they escaped censure while he was so violently attacked. Selden and the others argued that the law of nature is not reducible to self-interest, and that we always have a duty to put it into execution. As we shall see, for Hobbes its contents are deducible from the principle of self-preservation, and we need only put its precepts into practice when it is safe to do so.

3: THE LAWS OF NATURE AND JUSTICE

'This terme, the Law of Nature', wrote John Donne, 'is so variously and unconstantly deliver'd, as I confesse, I read it abundant tymes, before I understand it once'. 'Every man makes his own opinions to be laws of nature', said Taylor, 'if his persuasion be strong and violent'. The enquiry after 'a particular system of natural laws', he declared, 'hath caused many disputes in the world, and produced no certainty'.[38] Hobbes was not alone in trying to place talk about the law of nature on firm foundations. Suarez, Grotius, Selden, Taylor and others all attempted the same thing. Moreover, many of the general characteristics which Hobbes assigned to natural laws were conventional. What was new was his wholehearted attempt to derive them from the principle of self-preservation.

According to Hobbes, it is not possible to deduce the laws of nature from the actual practice of nations (EL1:15:1; DC2:1). Civil lawyers sometimes argued that the law of nature was common to animals and humans. According to Justinian's *Institutes* – a fundamental text of the Civil law – those moral norms which applied specifically to people were termed the law of nations, and were followed in all societies.[39] However, examination of the practices of nations suggested to many early modern Europeans that customs diverged drastically – a point underlined by accounts of recently discovered peoples in America and Asia. If the practice of nations was a reliable guide to universal moral truths, said Donne, 'How . . . shall we accuse Idolatry, or immolation of men to be sinnes against Nature?'. Spanish accounts of Hispaniola alleged that its inhabitants sacrificed 20,000 children each year. But the sacrifice of children is clearly immoral. So, Donne argued, moral truths cannot be deduced from the practice of all nations.[40] '[L]ewde and wicked custome', said Richard Hooker, could disguise even basic moral principles from multitudes of men.[41]

One way of circumventing this objection was to claim that the laws of nature could be deduced from the practices of most nations, or the most civilized nations, though not from the customs of absolutely all peoples. The civil lawyer Sir John Hayward equated the law of nature with 'the received custom, successively of al, and alwaies of most nations in the world',[42] and Grotius held that natural law could be derived 'with a very high degree of probability' from the practices of all or all the more civilized nations. He quoted Aristotle approvingly to the effect that 'the greatest proof of the truth of what we say is if all assent to it'. Nations whose customs were depraved, he argued, could simply be ignored, and he cited Andronicus of Rhodes' dictum that honey is in fact sweet, though sick men think it bitter.[43] 'The generall and perpetuall voyce of men', declared Hooker, 'is as the sentence of God him selfe'.[44]

But the attempt to read off laws of nature from the practices of most nations or the most civilized nations was itself open to serious objections. For one thing, there was the empirical point, raised by sceptics such as Montaigne and Charron, that the amount of agreement amongst nations on moral questions was in fact tiny. As we saw, Grotius rejected this sort of claim, asserting that there was substantial consensus at least amongst civilized nations. But others – such as Selden and Taylor – took the sceptical challenge more seriously. Taylor conceded that even such apparently civilized peoples as the ancient Spartans 'permitted natural injustices', for example encouraging their wives to have sexual intercourse with foreigners. So, he argued, Andronicus' dictum had little practical worth.[45]

Selden observed that nations often follow their own convenience rather than natural law, and he noted that their customs diverged even on such basic matters as homicide.[46] Hobbes likewise rejected the notion that natural law is deducible from the practice of nations. There was little point, he observed, in talking about 'the wisest and most civil nations' since 'it is not agreed upon, who shall judge which nations are the wisest' (EL1:15:1). But he claimed that there was one principle on which all people did in fact agree, namely that what is done in accordance with right reason is done by right. He concluded that right reason is the true guide to the law of nature, which he defined as 'the dictate of right reason, concerning what should be done or avoided continually to preserve life and limb, as much as in us lies' (DC2:1: 'Dictamen rectae rationis circa ea, quae agenda vel omittenda sunt ad vitae membrorumque conservationem, quantum fieri potest, diuturnam').

The idea that right reason tells us the contents of the law of nature was utterly conventional. Though the practices of nations might be a helpful guide to natural law, said Grotius, right reason was the ultimate ground for establishing that some practice was in fact according to the law of nature. 'The law of nature', he said, 'is the dictate of right reason, showing the moral necessity or moral baseness of any act according to its agreement or disagreement with rational nature'. In the opinion of Hooker, 'the lawes of well doing are the dictates of right reason'. The law of nature, said Donne, is 'but *Dictamen rectae rationis*' (the dictate of right reason). Suarez said much the same thing.[47] The notion was, indeed, a scholastic commonplace.

Later, Hobbes and Grotius were attacked by Roger Coke (grandson of the famous lawyer Sir Edward) for taking reason to mean the act of reasoning correctly from first principles and not the seat of reliable moral intuitions. Coke claimed that the laws of nature are self-evident. So no act of reasoning is required to grasp them.[48] In an annotation to *De Cive*, Hobbes observed that many people took 'right reason' ('recta ratio') to be 'an infallible faculty' ('facultatem infallibilem'), while he defined it as the act of reasoning by which conclusions are correctly drawn from true principles (DC2:1 Annot). Hobbes rejected the idea that people have intuitive knowledge of moral truths. The central axiom of his system – the principle of self-preservation – is not part of the law of nature since it is not deduced by reasoning. The notion that laws of nature may be deduced from first principles was not new with Hobbes; it was commonplace. Hobbes innovation was to argue that *all* the laws of nature are deduced in this way.

Hooker affirmed that the 'maine principles of reason are in themselves apparent' and therefore needed no rational demonstration. But the remoter conclusions of the law of nature were not so easily grasped, though Hooker assured his readers that 'there is nothing in it but anie man (having naturall perfection of wit, and ripeness of judgement) may by labour and travayle finde out'. Suarez argued similarly, asserting that no one could be ignorant of the first principles of natural law, but that ignorance of its less basic precepts was possible – and indeed explained how some nations had introduced laws which conflicted with that of nature. It was to be expected, he argued, that the common people especially would be ignorant of precepts which required a long chain of reasoning, most of all if ancient custom conflicted with the precepts in question.[49] For Hobbes, then, the laws of nature consisted in deductions from the axiom of self-preservation; for others, they included a variety of self-evident moral precepts, and the conclusions which could be deduced from them.

Hobbes described the laws of nature as 'Conclusions, or Theoremes concerning what conduceth to the conservation and defence' of ourselves. Viewed like this, he said, they were 'improperly' called laws. But as we have seen he added that they could also more accurately be termed laws inasmuch as they were 'delivered in the word of God, that by right commandeth all things' (Lev 15: 111/80). Hobbes claimed that the laws of nature were eternally binding (DC3:29) though he argued that some men through ignorance could be unaware of the existence of God (DC4:19; see also below, Chapter 6, section 1). So, in some sense at least, the laws applied independently of any knowledge of

God. This question of the grounds on which the laws of nature bind was much debated by medieval and early modern thinkers. There was nothing particularly startling about Hobbes' position.

According to the fourteenth-century philosopher Gregory of Rimini, any act that contravenes right reason is sinful independently of God or his wishes, 'for if God did not exist (which is impossible)' it would still be sinful to act against right reason.[50] Gregory's great contemporary William of Ockham, by contrast, argued that God's will alone defines good and evil, and the eminent theologian Jean Gerson likewise asserted the priority of God's will over right reason, claiming that actions are good or evil not because they are in accordance with or contrary to reason, but simply because God commands or prohibits them.[51]

Suarez attempted to steer a middle course between the two opposed viewpoints. He admitted that actions are good or evil in so far as they accord with or part from right reason, but argued that the law of nature was nonetheless a true law, instituted by God. Even if God did not exist, it would be sinful to act against right reason. In fact, God did exist and He commanded observance of natural law. Suarez claimed that his synthesis was the common teaching of theologians.[52] Certainly, his ideas were widely accepted in the seventeenth century. Hobbes had scathing things to say about Suarez's obscure language and contorted philosophy. But he also shared many of the Spaniard's views. Both men held that law is the command of a lawgiver, both stressed the distinction between counsel and command, and both insisted that every law requires a penalty. Moreover, both held that the dictates of right reason were binding independently of God's will, but that it was He who gave them the force of law.

Suarez asserted that the term justice has two distinct meanings. Firstly, he said, it can be taken as equivalent to all the virtues. Secondly, it can mean that particular virtue which is concerned with giving every man his due.[53] Hobbes used it in this second sense, describing justice as *'the constant Will of giving to every man his own'*, and observing that this was 'the ordinary definition of Justice in the Schooles' (Lev 15: 101/72; cf. DC epistle dedicatory, 9). It was, indeed, the standard scholastic definition, employed by Justinian's *Institutes*, by Aquinas, by Gerson and by a host of others.[54] Following Aristotle, theorists commonly divided justice into two kinds – commutative and distributive. Commutative justice was concerned especially with keeping contracts,[55] while distributive justice related to the equitable distribution of public goods or offices. The vice opposed to commutative justice was fraud, while that opposed to distributive justice was favouritism or *'Acception of persons'* (Lev 15: 108/77) (*'acceptio personarum'*: DC3:15). For example, Aquinas argued that a bishop who gave a benefice to an unqualified relative could be guilty of favouritism and therefore of flouting distributive justice.[56]

Hobbes' key innovation in the analysis of justice was to insist that it is always concerned with contracts. The 'definition of INJUSTICE, is no other than *the not Performance of Covenant'*, he said, while 'the observance of them' [i.e. covenants] is 'Justice' (Lev 15: 100–101/71–2; cf. DC3:3). As we shall see, the crucial point of this was to allow Hobbes to argue that sovereigns cannot commit injustice towards their subjects, and especially that the sovereign does not injure his subject if he takes his property without consent. But in most respects, the extent and significance of his innovation should not be exaggerated. Firstly, Hobbes – like previous thinkers – continued to hold that favouritism or 'acception of persons' is wrong, since it contravenes the natural law of equity (Lev 15: 108/77) which 'may be called (though improperly) Distributive Justice' (Lev 15: 105/75). Secondly, in claiming that justice strictly speaking is about keeping covenants Hobbes was in fact following a quite widespread tendency of moral theorists. 'The strongest thing which is against justice is to break a covenant', said the common lawyer Serjeant Saunders in Henry VIII's reign.[57] Amongst Hobbes' contemporaries, the Jesuit Paul Laymann argued that in a strict sense all justice is commutative, and Grotius said the same thing.[58]

In describing the laws of nature as principles reached by the exercise of right reason, and also as laws willed by God, Hobbes was occupying positions familiar to earlier writers. The same goes for his discussion of justice. What was original about Hobbes' treatment of natural law was not so much his analysis of its formal aspects as the *contents* which he assigned to it. And here his originality lay more in what he *omitted* than in what he included. Few would have denied that any or many of the laws which Hobbes listed were in fact laws of nature. Acts of cruelty and slander were commonly held to be breaches of natural law, as was favouritism and indeed drunkenness. Hobbes claimed that drunkards and gluttons acted wrongly since they destroyed or weakened the reasoning faculty without which no one could observe the laws of nature (DC3:25). This position was entirely conventional.[59] His critics attacked Hobbes' account of natural law less because of any inaccuracy in the laws he listed than because he omitted laws which they thought should be included.

As we have seen, Hobbes said that the laws of nature concerned 'what should be done or avoided continually to preserve life and limb'. He deduced the laws from the single principle of individual self-preservation. Others held that they were linked not only to the preservation of the individual but also to that of the community and the species, and they argued that the laws included a number of duties towards God. 'The main and principal laws of nature', said Bramhall, 'contain a man's duty to his God', and he observed that in his list of these laws Hobbes had 'not one word that concerneth religion, or that hath the least relation in the World to God'. A fundamental ground of Hobbes' errors, claimed Bramhall, lay in the fact that 'he maketh the only end of all the laws of nature to be the long conservation of a man's life and members'.[60]

This principle did indeed lead Hobbes into setting out what – for his contemporaries at least – were some odd moral conclusions. Most writers held that the Ten Commandments provided a summary of the moral law, which they equated with the law of nature. Commonly, the Commandments were divided into two groups – the first four, dealing with duties towards God, and known as the first table; and the remaining six, dealing with duties towards other people, and called the second table. A letter which Hobbes wrote in 1636 suggests that he then endorsed the usual view that all the Ten Commandments are 'the moral, that is, the eternal law' (EW7:454). But in *De Cive*, Hobbes specifically denied the status of natural laws to the first and fourth Commandments (which forbade having other gods and ordered that the sabbath be kept holy) (DC16:10). In *Leviathan* he went still further, claiming that all the Commandments of the first table 'were peculiar to the Israelites' and therefore not laws of nature (Lev 42: 357/282).

Again, and very importantly, though Hobbes held that the Commandments of the second table *were* natural laws (DC16:10), he argued that it was the civil law of each state which alone was to define what counted as breaches of these Commandments (DC6:16, 14:9, 17:10). Since there can be no property – no 'mine' and 'thine' – in the state of nature, he affirmed, it makes no sense to speak of my goods or my spouse in that state. Therefore, he concluded, theft and adultery are impossible there. So too is murder, for by the right of nature anyone may kill anyone else. What the biblical injunction against adultery really meant, said Hobbes, is 'avoid copulation which is forbidden by the laws' (DC14:9: 'Legibus vetitum concubitum fugies'). Again, the Commandment against murder means 'Do not kill the man whom the laws forbid you to kill' (DC14:9: 'Hominem quem vetant leges occidere, non occides'). All the Commandments of the second table, Hobbes argued, are similarly reducible to civil laws, and therefore – since it is the sovereign who makes law – to the sovereign's will. A major implication of this was that the sovereign has no obligation to consult his subjects before taking their property – since it is his will that defines what is their property. A main point of Hobbes' discussion of the Commandments – like his discussion of justice – was to vindicate taxation even without the consent of the taxed.

Virtually all of Hobbes' contemporaries held that the second table specified duties which applied to everyone, including sovereigns. The laws of particular states might add detailed provisions to these duties – say, on marriage or murder. But it was the laws of nature which set out the basic rules on such questions, for instance laying it down that killing was murder unless it was done in self-defence or in warfare, or as a punishment carried out on state authority. If the sovereign commanded a breach of these rules, he was to be disobeyed. While some held that the ruler was nonetheless never to be actively resisted, others permitted resistance in certain circumstances; but all agreed that the laws of nature set limits to the king's power.

A particularly sensitive topic in early seventeenth-century England was property. The first two Stuarts raised a number of levies without consulting their subjects in parliament. In the House of Commons, some argued that the king was not empowered to do this, since the subject held 'absolute property' in his lands and goods. The standard reply to this was that although the king would normally be unjustified in taking property without consent, he could do so in a case of necessity. However, many subjects were not convinced that any genuine necessity existed to justify such levies as the Forced Loan of 1626–7 and Ship Money in the 1630s (see Chapter 1, section 2 above). By arguing that subjects have no rights of property against the sovereign, Hobbes intended both to demolish the theory of 'absolute property' and to show that the king had no need to demonstrate necessity to justify his levies. The 'Kings word', he said 'is sufficient to take any thing from any Subject', and he declared that subjects ought not to 'ask whether his necessity be a sufficient title; nor whether he be judge of that necessity', but simply obey (Lev 20: 144/106; for Hobbes' doctrine on property see below Chapter 4, section 3). In reducing the contents of the laws of nature on theft, adultery and so on to nothing more than the ruler's will Hobbes deviated sharply from orthodoxy. He also greatly strengthened the sovereign's hand.

Another notable feature of Hobbes' morality is that he has little to say against sexual activities of however unusual a variety. For most moralists, to be licit sexual intercourse had to be calculated to propagate the species, but propagation has no moral value in Hobbes' system. As Tenison pointed out, Hobbes' theory does not suggest that there is anything wrong with the 'monstrous indecency' of 'buggery with a beast'.[61]

Hobbes' second (or, in *Leviathan* third) law of nature prescribes that people perform their covenants (DC2:1; Lev 15: 100/71). This was not a controversial principle. Everyone thought that people should, at least normally, fulfil their agreements. But the obligation to keep covenants occupies a crucial strategic position in Hobbes' theory. For covenants are the mechanism by which people are able to leave the state of nature and institute governments. Hobbes discussed covenants at length. Much of what he said about them was conventional, but he did introduce some important novelties into their analysis. The rules on covenanting which Hobbes put forward had little to do with the principle of self-preservation. That is to say, Hobbes nowhere showed that it would be rational for individuals keen to preserve themselves to introduce just these rules and no others. But the rules did serve the purpose of confirming that subjects can hold no property against the sovereign.

4: COVENANTS

In the English Civil War the idea that government rests upon the consent of the governed was commonplace amongst parliamentarians. Kings, they said, derive their power from the consent of the people, and are bound to rule within the limits which the people prescribed when they first transferred power to a monarch. They drew the conclusion that if the king exceeds these bounds, he (or at least his evil advisers) may be resisted. English royalists usually claimed that kings get their power directly from God and not from the people. So the people, they argued, could not resist their king.[62]

52   *Thomas Hobbes: Political Ideas in Historical Context*

They took Hobbes to task for basing his account of the origins of government upon consent. In France, some absolutists asserted that government did indeed begin with an act of consent by the governed, but held that this act could not be reversed and that it imposed no limitations upon royal power (see chapter 3, section 1 below). Hobbes likewise extracted absolutist conclusions from the premise that government begins with consent.

Hobbes argued that free and equal individuals in the state of nature will perceive that their situation is one of radical insecurity and will therefore be led to agree with each other to abandon the right of nature and erect a common power over themselves. He called mutual agreements contracts. Contracts, he said, might involve the immediate performance of what has been agreed to; on the other hand, one or all of the contracting parties might trust the others to perform in the future. Such contracts involving trust Hobbes termed covenants or pacts, and in Latin *'pacta'* (EL1:15:9; DC2:9; Lev 14: 94/66). It was sometimes said that pacts were so called because they led to *'pax'* or peace. This was certainly their function in Hobbes' system.[63]

According to Hobbes, if I covenant with some person to do or refrain from doing something and then later make a contradictory covenant with somebody else, this second covenant is void: a 'former Covenant, makes voyd a later' (Lev 14: 98/69). Again, children and madmen cannot make valid covenants, and any covenant to do something impossible is void (Lev 26: 187/140; 14: 97/69). These principles were standard. More controversial was Hobbes' assertion that no one can covenant away the right of self-defence. As we have seen, many royalists specifically denied this claim. Hobbes also argued that no one can be bound by contract to accuse himself. The idea that people cannot licitly be coerced into self-accusation was common, and in England it was frequently expressed by critics of the court of High Commission, the highest of the church courts. Procedure in this court included an oath – the so-called oath *ex officio* – which was tendered to defendants and by which they agreed to answer truthfully any questions that might be put to them. But in answering the questions they might accuse themselves. The ban on self-accusation was later enshrined in the fifth amendment to the American constitution.[64]

Hobbes argued that valid contracts must be accepted by both parties, and this was uncontroversial. But he drew the unusual and highly important conclusion that without some special revelation, we cannot covenant with God except through the mediation of the civil sovereign, God's agent on earth. For we cannot be sure that God accepts what we promise (EL1:15:11; DC2:12; Lev 14: 97/69). Both Protestants and Catholics agreed that Scripture makes clear what promises God accepts, though they disagreed on just what the relevant activities are (for example, Protestants rejected vows of celibacy). So Hobbes' argument was strikingly original. It also had an immediate practical application, for in 1638 the Scots entered into a covenant with God to oppose royal policy – and it was the Scots' rebellion which led to the collapse of Charles I's regime in England.[65] More generally, many Protestants were in the habit of making personal covenants with God, and Catholics took vows to Him. Hobbes thought the tendency of this was to limit obedience to the king.

A long-standing bone of contention in the analysis of covenants was the question of *'nuda pacta'* or naked pacts. This term was used in more than one sense. Often a naked pact was taken to be an agreement in which one party undertook to give something to the other but received no recompense or 'consideration' in return. The well-known early-sixteenth century common lawyer Christopher St German held that such a 'nude contracte . . . ys voyde in the lawe and conscience'.[66] In his *Parallele* of the civil (or Roman) law, the canon law (which was used in church courts) and the English common law, the Elizabethan lawyer William Fulbecke observed that common law requires 'consideration' to validate a contract, but that 'by the Canon Law *nudum pactum* doth bind'. The sixteenth-century French civil lawyer François Connan argued that a promise is not binding if the promisor receives no consideration. This was Hobbes' view, though he did not term such promises naked pacts (DC2:8). Grotius, by contrast, argued that promises (unlike mere statements of intention) bind even without consideration, and he cited civil law texts in support of his opinion. Aquinas, the casuist Thomas Sanchez, and others likewise argued that promises are binding without consideration.[67]

A second definition of a naked pact held that it was one which was not confirmed by an oath or legal formality.[68] Hobbes himself used the term in this sense, arguing that 'a naked pact obliges no less than one to which we have sworn' (DC2:22: *'pactum* nudum non minus obligare, quam in quod iurauimus'). On this point his opinion was the same as that of Grotius and Aquinas. Grotius attacked Connan's idea that what gives contracts obliging force is human law, asserting that human law-makers are themselves instituted by contract and therefore that the obligation to obey man-made laws necessarily presupposes a prior obligation to keep contracts.[69] In Hobbes' system, too, the obligation to keep covenants stems from the law of nature, which is prior to all man-made laws.

The idea was utterly conventional, and for most thinkers unproblematical. But from the point of view of Hobbes' theory, the notion raised important questions connected with his account of the state of nature. If *all* covenants are binding in the state of nature, then it is possible for people there to institute property and adopt rules to regulate their lives. Hobbes, however, asserted that the state of nature was anarchic and that no property existed there. Again, if *no* binding covenant can be made in the state of nature, then it will be impossible to leave that state. At this point in his analysis, Hobbes turned to the concept of 'just fear' which featured largely in the writings of lawyers and casuists.

If two people in the state of nature covenant with one another to perform some action in the future, said Hobbes, and if a 'just fear' (*justus metus*) arises in either that the other will not in fact perform, then the covenant becomes void (DC2:11; cf. 13:7, EL1:15:10; Lev 14: 96/68). In other words, if I have a 'just fear' that once I have fulfilled my part of the bargain you will break yours, then it would be irrational for me to keep the covenant, for by doing so I would give you an advantage in the war of all against all (Lev 14: 96/68). An obvious question here is: what does Hobbes mean by 'just fear'? If we adopt Hobbes' own definition of justice, we might be led to conclude that my fear is just if I have undertaken by contract to be afraid, or at least if I have made no contractual commitment to be fearless. But this is not at all what Hobbes means by the term. Rather, he uses 'just fear' to denote a reasonable, well-grounded fear, which would absolve people from obligations they would ordinarily have had. The concept was employed in exactly the same way by casuists and lawyers, indicating the kind of fear which would seriously and reasonably affect a man of firm character, and which might excuse otherwise sinful actions performed as a consequence of the fear. Sarpi, for instance, argued that 'just fear' excuses us for disobeying human and even divine positive laws, though not the law of nature. A typical formula was 'a grave and just fear, and one which may affect a man of firm character'. What Hobbes did was to adapt a traditional idea to his own purposes.[70]

The same goes for his argument that covenants made under

coercion bind. 'It is commonly asked', he said, 'whether pacts extorted by fear are binding or not' (DC2:16: 'Quaeri solet an *pacta* ea quae metu extorquentur, obligatoria sint necne'; cf. EL1:15:13). He answered the question in the affirmative, and gave some examples to illustrate his point. One was of a man who is captured by a thief and to save his life promises to pay the robber a thousand pieces of gold on the next day (DC2:16). Of course, covenants are free acts. But if a man acts under fear can he be said to be free? Yes, said Hobbes, for 'Feare and Liberty are consistent; as when a man throweth his goods into the Sea for *feare* the ship should sink, he doth it neverthelesse very willingly, and may refuse to doe it if he will' (Lev 21: 146/108).[71]

Hobbes was quite correct in suggesting that discussion of these problems was common. Casuists did indeed frequently confront the problem of extorted contracts. Moreover, they commonly used both of Hobbes' examples and took the same line as he did. Some thinkers, indeed, argued that contracts made under coercion are void. A promise extracted by force is void, said Cicero – and Ames, Fulbecke and Filmer took the same line.[72] But others disagreed. According to the Spanish Jesuit Azorius, the common view of casuists and theologians was that 'he who promises on oath to give money to a thief in order to avoid being killed by him, is bound to keep his oath and fulfil his promise'. He spelled out that this applies to simple promises as well as to those accompanied by oaths.[73] Promises and contracts extorted by fear are valid according to the law of nature, said Laymann and Sanchez. Laymann and others argued that actions performed under coercion are voluntary (or, strictly, more voluntary than involuntary) and cited the case of the man who throws his goods overboard to prevent shipwreck.[74] Grotius asserted that 'he who promises anything through fear is obliged' to perform what he has promised, and he too used the example of the man who throws his goods overboard – an example derived from Aristotle. Amongst Englishmen, Sanderson and Taylor argued similarly.[75]

So the arguments on just fear and on covenants made under coercion were variations upon familiar themes. Hobbes said little to show that self-preservation requires these particular rules. Arguably individuals bent on preserving themselves would in fact adopt rather different principles.[76] But these rules served major functions in his theory. By adopting the unusual argument that covenants are invalidated if just fear of non-performance arises, Hobbes was able to underpin his claim that property is impossible in the state of nature – for this kind of just fear is extremely likely to occur there, since no coercive power exists to enforce contracts. Hobbes concluded that the subject's rights of property are derived from the will of the sovereign. It followed that sovereigns need never obtain their subjects' consent to taxation – and this was one of the major practical conclusions of Hobbes' political writings. So Charles I's extra-parliamentary levies – such as the Forced Loan and Ship Money – were fully justified.

The principle that covenants made in conditions of fear are valid was also of great importance in Hobbes' system. Without it the inhabitants of the state of nature would be unable to institute a commonwealth. For they live in a situation of constant fear, and therefore any covenants they make, including the covenant instituting government, are made under fear. Again, he argued that the other two possible original forms of government besides the institutive – namely patriarchal government and government by conquest – likewise arise from covenants made under fear. These various covenants are the subject of the Chapter 3.

3
The Origins of Government and the Nature of Political Obligation

1: THE INSTITUTION OF A COMMONWEALTH

Hobbes held that the earliest states had been families. 'It is evident', he said, 'that the beginning of all Dominion amongst Men was in Families', and that the father 'was absolute Lord of his Wife and Children' (Dialogue 159). He also thought that commonwealths could begin through conquest. But he spent most space on the commonwealth by institution, which he treated as the paradigm case of how government arises. The same emphasis on commonwealths by institution was characteristic of most contractualists or contractarians – that is to say, of thinkers who grounded government upon the consent of the governed. Many such theorists drew the conclusion that rulers are now bound by the conditions of the original contract and may be resisted if they breach them. Among medieval Roman lawyers, however, it was commonly argued that the original contract was an absolute transference of power from the people to its ruler. A number of Hobbes' French contemporaries – including Cardin le Bret and Edmond Richer – likewise argued that the people originally consented to absolute monarchy, and that kings were irresistible. Hobbes similarly claimed that the people's consent instituted absolute sovereignty. But he added a novel argument, asserting that government by institution arises through a contract between individuals excluding the sovereign, and not between the people and its ruler. Since the sovereign is not a party to the original contract, he said, he is plainly incapable of breaking it (EL2:2:2–3; DC7:14; Lev 18:122/89).[1]

Virtually all contractualists argued that by nature individuals possess no power to punish, but only a power to defend themselves against attack. The power of punishing criminals, they said, first arises when individuals or families congregate together in a community large enough to provide for its internal welfare and external defence. In their opinion, human nature required that people join together in such a 'perfect community' (*communitas perfecta*). Since God was regarded as the author of nature, the powers which the community possessed to punish criminals – and to make laws specifying what was punishable – were held to stem from God.

Hobbes took a very different line, claiming that the powers of the sovereign are precisely the same as those of the individual in the state of nature. This was not wholly original. Grotius had similarly asserted that the power of punishing was vested in individuals before the institution of states. But the Grotian – and Hobbesian – line was distinctly uncommon. So Hobbes was an unusual contractualist in two major respects. Firstly, he drew absolutist conclusions from consensual premises, and did this by means of denying that the sovereign was a party to the original contract. Secondly, he held that the sovereign's right is identical with the right of the individual in the state of nature.[2]

According to Hobbes, the attainment of peace requires that people lay down the right of nature, for it is because of their attempts to exercise this right that the state of nature is so insecure. It would be irrational for just a few people to lay down their right, since by doing so they would simply make it easier for others to overcome them. What is needed to bring peace is that all the individuals concerned agree together that each will lay down his right provided that all the rest do so too. The only person exempted from this agreement is the future sovereign, who can be any individual or individuals.

Since the sovereign does not hand away the right of nature, the people who do so can add nothing to the right he already possesses. They *transfer* their right to him, says Hobbes, not by giving him any new right but by agreeing to stand out of his way as he exercises the old right (Lev 14: 92/65). The sovereign's right of nature is a right to do all he thinks necessary in order to preserve himself. Self-preservation requires that we seek and maintain peace. So the right of nature empowers the sovereign to do whatever he conceives to be necessary to preserve peace. Subjects, says Hobbes, also undertake not to assist those whom the sovereign punishes. Moreover, the end or purpose of the covenant, though not necessarily its express words, requires that subjects aid the sovereign by performing dangerous and dishonourable offices, or even by risking their lives as soldiers, when the preservation of the peace demands such a course (Lev 21: 151–2/112).[3]

Hobbes argued that to set up a commonwealth, people in the state of nature must assemble, and agree upon a sovereign to whom they will transfer the right of nature. In *De Cive* and the *Elements* he claimed that this original assembly of the people was itself the sovereign; for the decisions of the assembly bound everyone, and it united the wills of all into a single will (DC6:1; EL2:1:2–3). This implied that the original form of government in every institutive commonwealth was democracy – an implication which Hobbes in fact embraced in *De Cive*: 'those who come together to set up a city, are almost by that very act of coming together a democracy' (DC7:5: 'Qui coierunt ad civitatem erigendam, pene eo ipso quod coierunt, *Democratia* sunt'; cf. EL2:2:1).

From the perspective of royalist polemic, this assertion was very awkward. For the doctrine that governments were originally democratic featured largely in the theories of critics of absolute monarchy, including parliamentarian propagandists of the Civil War period. They used the idea to argue that the powers of kings must at first have been derived from the sovereign people, and that it was open to the people when they appointed their ruler to impose conditions or limitations upon his power. Some concluded that the king is the people's delegate, and that they may call him to account at will. The maxim that the king is superior to each individual subject, but inferior to the whole commonwealth – *singulis major, universis minor* – was frequently repeated in parliamentarian pamphlets.[4] Hobbes mentions – and of course vigorously rejects – the idea in *Leviathan* (Lev 18: 128/93). Earlier, some Catholic thinkers (like Azorius and Suarez) had adopted a slightly different position, arguing that the king was ordinarily superior even to the community considered as a whole, but that he ceased to be so if he became a tyrant. In their opinion it was the people acting in consultation with the pope who were to judge whether the king was a tyrant. So, effectively the king could always be adjudged a tyrant.[5]

Hobbes rejected the argument that if government was at first democratic, kings are now accountable to the people. A commonwealth, he claimed, must have a single will which is the sovereign's, for otherwise it will not be a united body, and disunion spells anarchy. So if a king *is* accountable to, say, parliament, then either there is no commonwealth, but disunion, or it is parliament which is sovereign. It follows that if a king is exercising sovereign power, he is not accountable to parliament, and this is one of the main practical messages of the *Elements* and *De Cive*. Before 1640 it was widely agreed that Charles I did hold sovereign power in England, though some thought that at least in certain circumstances he was accountable to parliament for how he exercised that power. Hobbes' argument is that such thinking is woolly, for sovereigns are accountable only to God.

From 1642 onwards, however, the question of who held sovereignty in England became increasingly contentious. Parliamentarians put forward claims that the two Houses were sovereign or at least shared in sovereignty, and the Houses in fact performed such sovereign acts as levying taxes and raising troops. They used the idea of original popular sovereignty to justify such actions. For if the whole community – and therefore its representative assembly, parliament – had originally been sovereign, then unless there was firm historical evidence that they had at some point resigned their sovereignty – and of course they argued that there was no such evidence – the presumption was that they were now still sovereign.

Most royalists rejected the idea of original popular sovereignty. In *Leviathan* Hobbes at last did the same thing – and the means he employed was the doctrine of authorisation.[6] He argued that what gives the commonwealth its unity is the sovereign, who has been authorised by each individual to bear his person. That is to say, every individual authorises all the actions of the sovereign. He therefore becomes the agent or representer of each particular person, and the group acquires unity because all its members have the same representer. 'A Multitude of men', Hobbes declared, 'are made *One* Person, when they are by one man, or one Person [i.e. an assembly acting with a single will] Represented; so that it be done with the consent of every one of that Multitude in particular. For it is the *Unity* of the Representer, not the *Unity* of the Represented, that maketh the Person *One*' (Lev 16: 114/82). If the representer is one man, then the form of government is monarchy, and therefore it is not necessary that the original form of government in every commonwealth be democratic.

Hobbes claimed that if individuals give their representer 'Authority without stint' (Lev 16: 114/82) then whatever he does is the action of each individual. The idea can be traced back to medieval canon law. Canonists argued that if a group gave 'full power' to its representer, then it was obliged by his acts even if it had not specifically consented to them in advance. The same idea was incorporated in English writs of summons to parliament from the thirteenth century onwards.[7] Those elected to the House of Commons, said the writ of Hobbes' day, are to have 'full and sufficient power' to consent to what is done in parliament.[8] The purpose of this was to ensure that members did not try to retract their consent to taxation and legislation by later claiming that they had exceeded the powers granted to them by their constituents.

In the Civil War, similar ideas were employed by parliamentarians to show that parliament is not accountable to the people, and indeed that parliament *is* the people. Thus Henry Parker declared that 'the Parliament is indeed nothing else, but the very people it self artificially congregated or reduced by an orderly election, and representation, into such a Senate or proportionable body'. 'The parliament', he said, 'is indeed the State itself'. For it was only in parliament that the people became a united body. Considered apart from parliament, the people was a cumbersome or 'moliminous' body, whose motions were unregulated. So for Parker, the idea that the people could call parliament to account if it acted wickedly was based on confused thinking. Parliament was the representative of the people in the sense that the disunited multitude granted it full power, which it therefore could not retract. On similar grounds, Charles Herle likewise argued that parliament was irresistible.[9]

These authors claimed that *parliament* is the people's representative, and indeed that it *is* the people. Hobbes made the same claims for the sovereign. He noted with amazement that amongst the English a king whose direct ancestors had ruled for six hundred years 'was notwithstanding never considered as their Representative' (Lev 19: 130/95). The absolutist equation between king and commonwealth was, indeed, sometimes drawn by English writers. 'Every prince', said the cleric Samuel Collins, 'is virtually a whole kingdom'.[10] But Hobbes was right in claiming that the king was hardly ever styled the people's representative – and there was good reason for this. English absolutists did not derive the king's power from the people, preferring to regard the monarch as God's representative. Opponents of absolutism were unlikely to argue that the king represents the people, since they looked to another representative – parliament – to limit the king's power.

But the notion that the sovereign represents and indeed *is* the people made perfectly good sense from the perspective of Hobbes' theory, which was at once contractualist and absolutist. It also enabled him to reduce to nonsense the claim that kings are inferior to the whole community. He considered the contention that kings are superior to 'every one of their Subjects' (considered as individuals), but *'Universis minores,* of lesse power than' their subjects considered 'all together', or as a collective entity. It was, he said, the sovereign who bore the person of this collective body. To say that he was inferior to the commonwealth was therefore to say that he was inferior to himself, and this was manifestly absurd (Lev 18: 128/93).

It was similar reasoning which led Parker and others to argue that parliament *is* the people, and so accountable to no one. They asserted that parliament's powers are held unconditionally. Hobbes himself observed that theorists did not derive the powers of popular assemblies from a conditional grant by the people. '[W]hen an Assembly of men is made Soveraigne', he asserted, 'then no man imagineth any such Covenant to have past in the Institution; for no man is so dull as to say, for example, the People of *Rome,* made a Covenant with the Romans, to hold the Soveraignty on such or such conditions; which not performed the Romans might lawfully depose the Roman People' (Lev 18: 123/90). Where popular assemblies rule, Hobbes argued, they *are* the people, and where a king rules he *is* the people (EL2:2:11; DC12:8).

As we have seen, the idea that parliament is the people was expressed by such writers as Parker and Herle, and has roots which go far further back; but it was *not* universally accepted. From the early 1640s men began to claim that the people could call parliament to account if it acted evilly, and after 1645 the Levellers in particular vigorously championed the notion that parliament did hold its powers conditionally.[11] To paraphrase Hobbes, such thinkers argued that the people of England *had* made a covenant with the English as represented in parliament to hold sovereignty conditionally. Since parliament had failed to abide by the conditions imposed upon it, their argument continued, the people were justified in resuming power. Ideas like these played a crucial role not only in Leveller propaganda, but also in justifications of the army's imposition of its will upon parliament, culminating in Pride's Purge of December 1648 and the execution of the king in the following month. Some men, then, *were* so dull as to say that the sovereign assembly is accountable to the people. Hobbes ignored this. But he by no means ignored all the debates to which the Civil Wars gave rise, for his doctrine on conquest was applied in *Leviathan* to make telling points directly relevant to current debate on the Engagement of 1650.

2: CONQUEST

Hobbes held that commonwealths can begin not only by institution but also by conquest. He claimed that William of Normandy acquired the crown of England in this manner (Lev 24: 172/128). The conqueror became sovereign, he argued, when the defeated population consented to his rule. Conquest, he claimed, was reducible to covenant. Despotical dominion – dominion by conquest – 'is then acquired to the Victor', he said, 'when the Vanquished, to avoyd the present stroke of death, covenanteth either in expresse words, or by other sufficient signs of the Will, that so long as his life, and the liberty of his body is allowed him, the Victor shall have the use thereof, at his pleasure' (Lev 20: 141/103–4). 'It is not therefore the Victory, that giveth the right of Dominion over the Vanquished', he concluded, 'but his own Covenant' (Lev 20: 141/104). Until the covenant was made, the defeated person was the victor's slave. He might obey him to avoid death or physical ill-treatment, but since no covenant had been made between them, he was under no *obligation* to obey: slaves 'have no obligation at all; but may break their bonds, or the prison; and kill, or carry away captive their Master, justly' (Lev 20: 141/104).

Once the covenant has been made, the defeated individual ceases to be a slave and becomes a servant, while the victor becomes a master. The relationship between master and servant, Hobbes argued, is identical with that between sovereign and subject in a commonwealth by institution. Since the victor does in fact fulfil his part of the covenant by sparing the life of the vanquished, Hobbes' rules on covenanting entail that the defeated person is obliged to perform the covenant by obeying. However, there are limits to this duty of obedience. As always, the servant retains the right of self-defence. Moreover, if the master ceases to preserve the life of the servant by failing to protect him, the covenant becomes void. Thus, if the servant is captured by a later invader, the old covenant lapses and he may make another with the new conqueror (DC8:9).[12]

The idea that William I conquered England was sometimes expressed by royalists; examples are James I, and Hadrian Saravia – a trenchantly absolutist thinker, whose work was praised by Grotius.

Like Hobbes, they placed little emphasis on William's conquest – and for the same reason.[13] They argued that all true kings possess sovereignty, which is absolute and indivisible: conquerors, they claimed, held no more powers than other kings. Such thinkers were happy enough to admit that conquest was a possible – perhaps even a common – means of acquiring kingdoms. But since they argued that all kings possess the same powers, and derive these powers from God alone, conquest played little part in their theory. It could, however, serve a useful ancillary function, for if a government began by conquest then those who tried to trace its origins to popular consent had arguably got their history wrong.

In the Civil War, Henry Ferne claimed that in England and many other countries kings had been established by conquest. His point in doing this was to rebut the parliamentarian notion that all government begins with an act of consent by the people. Royalists usually said that the power of governing comes from God alone. Only God possesses the power of life and death over individuals. So whoever now holds such power here on earth must have drawn it directly from Him, and not from popular consent. From the standard royalist perspective, therefore, the consent of the people *could* not transfer political power to a sovereign. But it was still pleasant to be able to show that parliamentarian claims were not only philosophically but also historically false – in that monarchy in England could be traced back to conquest, and not to consent.[14]

Those who argued against absolute monarchy commonly grounded their case on the idea that all rulers derive their power from the consent of the subject – or at least that the ruler of their own kingdom did so. Thus, some English opponents of absolutism claimed that William I had ruled not as a conqueror but as the rightful heir to the throne, or as the people's appointee. Another approach was to claim that *all* conquerors require popular consent to legitimate their titles. Suarez admitted that kingdoms may be acquired by victory in war, but argued that this was ultimately reducible to contract, or to something close to contract. If the war was unjust, he asserted, then the victor acquired no rights except through the consent of the people. On the other hand, if the war was just then the conqueror gained dominion over them as a 'just punishment' of them. Just punishment, he claimed, had the force of contract in transferring power. Again, said Suarez, to wage a just war a man must already hold political authority – for making war is a sovereign act. Before a man can add to his dominions by lawful conquest he must already hold political authority, and as this plainly cannot stem from conquest, it must stem from consent. Conquest therefore can at most explain how a king extends his dominions, not how he becomes a king in the first place.[15]

Similar ideas circulated in England long before the Civil War. Sir Edwin Sandys argued in 1614 that unless the people consented to the conqueror's rule his government was unlawful and he could be removed. A little later the Jesuit John Floyd claimed that 'lawfull Conquest' did not make the conqueror a king until the vanquished consented. 'Yf they refuse to yield', Floyd argued, 'he hath the right of the sword to force them, not the right of the Prince to governe them, till they consent'; but this 'consent being yielded, then there begins a new Society and Commonwealth compacted of Conquerors and the people conquered, and the Prince of the conquering side becomes Kinge'. Precisely the same line was commonplace in parliamentarian propaganda during the Civil War. Mere conquest, argued the Independent cleric Jeremiah Burroughes, confers no political rights: 'that which subjects my conscience to such a one, is the submission upon some compact, covenant or agreement'. What the conqueror held 'merely by force, without any consent and agreement, was no power, no authority at all, but might be resisted'. Conquest, said the parliamentarian lawyer John Cook, confers no lawful title unless the people subsequently consent: until 'the people consent and voluntarily submit to a Government, they are but Slaves, & in reason they may free themselves if they can'. Patently, these ideas are very close to those of Hobbes.[16]

For Floyd, Burroughes, Cook and similar theorists a major point in arguing that only consent validates a conqueror's claims to political authority was to show that even in states where the monarch was a conqueror – or the descendant of a conqueror who had inherited his ancestor's power – his authority was derived from popular consent and limited by whatever conditions the original contract specified. The agreement that made the victor king, said Floyd, required him to rule according to 'lawes and conditions agreed upon: which conditions if he neglect, he is no less Subiect and corrigible by the Commonwealth, then Kings made by election'.[17] So kings by conquest are bound by contractual conditions. There is an obvious practical problem here. Since the conqueror has the sword he can use it to encourage the people to come to an agreement which is as advantageous as possible to himself. All conquerors, it seems, are in a position to coerce the conquered population into covenanting away their rights, and pledging absolute obedience. Conquerors, then, can force the conquered population into making them absolute rulers. But if this is so, the whole purpose – for anti-absolutists – of claiming that conquerors rule by consent has failed.

Two means were explored of evading this difficulty. One was to suggest that consent must be uncoerced if it is to create legitimate authority. The other was to claim that people have an extensive set of rights which they cannot give away. Hobbes rejected both of these notions. Covenants made under coercion, he insisted, are valid, and people have few inalienable rights except that of personal self-defence. Such thinkers as Grotius, the civil lawyer Sir John Hayward, and the cleric Saravia similarly argued that just as an individual could give or sell himself into servitude, so could a whole people. As 'a private man', said Hayward, 'may altogether abandon his free estate, and subject himself to servile condition, so may a multitude passe away both their authoritie and their libertie by publike consent'. Grotius held that the victor in a just war gains as much power as a ruler to whom a people fully subjects itself.[18] Of course, in Hobbes' theory there is neither justice nor injustice in a state of war, and so all conquerors have the same power. Grotius not wholly convincingly gave the people certain residual rights of resistance against their sovereign, while Hobbes (and Hayward and Saravia) granted them none.

Hobbes' theory of conquest incorporated many familiar features. It was fully worked out in the *Elements* and *De Cive*. In *Leviathan*, Hobbes drew some important practical applications from the theory, but added little to it. Already in the *Elements* he asserted that 'if a commonwealth is overthrown', and the individual 'yieldeth himself captive' to the conqueror, he thereby acquires an obligation to obey him. Since 'no man can serve two masters' his duties to his former sovereign automatically lapse (EL2:2:15). In *De Cive*, Hobbes repeated the argument (DC7:18).

By the time he completed *Leviathan*, the whole question had become highly topical. For Charles I had been defeated in war and executed, while his son Charles II seemed unlikely to make good his claims to the English crown. In January 1650 the new government of the Commonwealth required that all males aged over eighteen subscribe to an Engagement, which ran as follows: 'I do declare and promise, That I will be true and faithful to the Commonwealth of England, as it is now established, without a king or House of Lords'.[19] The men who introduced this measure were much influenced by puritan thinking, and did not lightly coerce their fellows into taking oaths. The Engagement did not invoke God and so was no oath, though modern commentators on Hobbes almost invariably describe it as one. It did, however, raise problems which were connected with oaths. Many of those who were asked to take the Engagement had earlier taken the oaths of allegiance and supremacy, and some had subscribed to the Solemn League and Covenant of 1643. In short, many people who had pledged allegiance to Charles I and his heirs were now being asked to do the same to an entirely different government. This sparked off a major pamphlet war in the opening years of the 1650s.

Unquestionably, Hobbes was aware of the debate on the Engagement, and a number of passages in *Leviathan* are clear contributions to the controversy. 'The Obligation of Subjects to the Soveraign', he declared, 'is understood to last as long, and no longer, than the power lasteth, by which he is able to protect them'. If your sovereign is unable to protect you, then you recover the natural right to protect yourself, for the 'end of Obedience is Protection; which, wheresoever a man seeth it, either in his own, or in anothers sword, Nature applyeth his obedience to it, and his endeavour to maintaine it' (Lev 21: 153/114).

His reading of 'divers English Books lately printed' made it plain to Hobbes 'that the Civill warres have not yet sufficiently taught men, in what point of time it is, that a Subject becomes obliged to the Conquerour; nor what is Conquest; nor how it comes about, that it obliges men to obey his Laws'. When 'the means of his life is within the Guards and Garrisons of the Enemy', he asserted, a man has 'liberty to submit' to him, for it is then from that enemy and not from his former sovereign that he receives protection. Moreover, he observed, as things stood people who submitted to the new government would be allowed to retain their estates, while those who refused to submit would lose them. So you would actually assist the new government more by refusing to submit than by submitting (Lev, 'Review and Conclusion', 484–5/390). A few years later, Hobbes claimed that *Leviathan* 'hath framed the minds of a thousand gentlemen to a conscientious obedience to present government, which otherwise would have wavered in that point' (*Six lessons to the professors of the mathematics*, EW7:336). At the very end of *Leviathan*, he declared that the work had no 'other designe, than to set before mens eyes the mutuall Relation between Protection and Obedience' (Lev, 'Review and Conclusion', 491/395–6).

The emphasis on 'the mutuall Relation between Protection and Obedience' of *Leviathan* is unparalleled in the *Elements* and *De Cive*. The very same emphasis is present, however, in the writings of a number of defenders of the Engagement of 1650. Such writers as Nedham, Wither and Osborne used the notion of a mutual link between protection and obedience to argue that since the government of the Rump was in fact protecting the English population, it now had a claim to their obedience.[20] We might be tempted to conclude that this argument bears witness to a new and pragmatic approach to politics which first arose amongst the defenders of the Engagement, and to which Hobbes gave particularly clear expression. But it is difficult to clinch this argument, for there was in fact nothing particularly novel about the assertion that protection and obedience are mutually linked – nor, indeed, about most of the other contentions made by those who supported the Engagement. The connection between protection and obedience was frequently drawn in the 1640s.

According to the political commentator Clement Walker, the leading parliamentarian Henry Ireton argued in 1647 that by refusing to assent to some wholesome laws the king had effectively 'denyed safety and protection to his people' and 'this being denied they might wel deny any more subjection to him, and settle the Kingdom without him'. At Putney earlier that year, Ireton argued that anyone who was actually protected by the laws of the land was bound to obey those laws, whether or not he had consented to them. In 1646 Thomas Chaloner likewise affirmed that anyone who enjoyed the protection of English laws automatically owed obedience to them: 'hee being protected by the Lawes of England, hee becomes thereby subject to those Lawes, it being most certaine that *protectio trahit subjectionem, & subjectio protectionem*' [protection draws with it subjection, and subjection protection]. '[N]o man', he declared, 'can be said to be protected that is not withall thereby subjected'.[21]

In 1645 William Ball asserted that 'it is an Axiom Politicall, *ubi nulla protectio, ibi nulla subjectio*' [where there is no protection, there is no subjection], and he used the axiom to argue that if either king or parliament failed to protect them, the people could provide for their own safety. In the opening year of the Civil War, John March claimed that 'just as the subject is bound to obey the king, so the king is bound to protect the subject'. '[I]t is truly said', he remarked, 'that . . . Protection draweth subjection, and subjection protection'.[22]

Patently, the defenders of the Engagement were doing nothing novel in asserting a mutual link between protection and obedience – and that also, of course, goes for Hobbes. It goes, too, for the writers of the 1640s. Much the same viewpoint was expressed in a famous test case on allegiance of 1608 – Calvin's Case. Reporting the case, Sir Edward Coke remarked that 'as the subject oweth to the King his true and faithful ligeance and obedience, so the Soveraign is to govern and protect his Subjects'. '[I]t is truly said', he added, 'that *protectio trahit subjectionem, & subjectio protectionem*' [protection draws with it subjection, and subjection protection]. If an alien comes to England, he affirmed, then 'he is within the king's protection; therefore so long as he is there, he oweth unto the King a local obedience'. The subject's allegiance, he declared, is 'of as great an extent and latitude, as the royal power and protection of the King, and *e converso*' [i.e. vice versa]. The king of England still laid claim to France, but no one born there since he had lost 'actual possession' of it was subject to him. Coke argued that the king did indeed have an 'absolute right' to French territory, but this right had little meaning, for he held that only the inhabitants of lands in his actual possession were subject to the monarch. Subjection, then, was linked to protection, and not to the lawfulness of the king's title.[23] Other pre-Civil War thinkers argued similarly. 'Nature teaches him to obey him that can preserve him', wrote John Donne. Usurpers held lawful authority, though they did not acquire it legitimately, said the cleric William Sclater. The doctrine that once a usurping government had been 'throughly settled' it gained legitimacy – a doctrine very close to Hobbes' own – was enshrined in the canons of 1606, which were the official pronouncements of England's clergy. James I took exception to the doctrine, which implied that he could lose royal power if there was a successful rebellion against him, but the fact that the clergy adopted this position suggests that it was widely accepted.[24]

The problem of what, if anything, legitimates the rule of a usurper caused difficulties for those who – like most royalists – argued that the power of governing is not derived from the consent of the governed. To set down rules legitimating usurpation was, as James I saw, to invite rebellion since the rebels could acquire lawful power by fulfilling the rules. Of course, it was possible to construct rules to which it was difficult to adhere. Some argued that three generations or a hundred years were required to render the usurper's line legitimate, while Ferne claimed that the rule of an unjust conqueror becomes legitimate only in his successor, for 'in the succession . . . that providence which translates kingdoms manifests itself'. It was, however, difficult to mount a convincing case to show that any particular length of time was needed to legitimate usurpation, and still more difficult to prove that practice accorded with theory on this point. In England the Tudor and therefore the Stuart dynasty was arguably founded upon usurpation, though subjects of course rarely questioned the legitimacy of their rulers' titles – at least in public. Not surprisingly, most thinkers chose not to dwell on these problems. William Dickinson was typical in arguing that 'it is not for those whom God hath appointed to obey, to examine titles and pedigrees, or how kings came to their power, and to be rulers over them'. 'It sufficeth', he declared, 'that being under, we must obey'.[25]

Once Charles I was defeated and executed, this line of argument could be and was used to vindicate obedience to the Rump. Royalists such as Sanderson and Filmer now claimed that usurpation can never lead to a legitimate title. The subject's duty of allegiance, said Sanderson, is 'perpetual and indispensable'. In circumstances where force prevented us from obeying the true ruler, we were excused from doing so, but we could never acknowledge a usurper as lawful. Filmer argued similarly.[26] An awkward implication of the theory was that if our ruler's title is itself ultimately derived from usurpation – since he descends from an unjust conqueror or rebel – we are obliged to withhold allegiance from him and pay it instead to the descendant of whoever was king before the usurpation. Indefeasible hereditary right was not a very practical doctrine, and it had not been royalist orthodoxy before the Engagement controversy. On the other hand, the notion of a mutual link between protection and obedience was well-known before 1650. Arguably, Hobbes' doctrine on this question was one of the least innovatory parts of his system. By contrast, his teaching on the family was deliberately and even provocatively novel.

3: THE FAMILY AND THE STATE[27]

Virtually all of Hobbes' contemporaries held that by nature fathers have authority over their families, and that this authority is derived from God and not from their children. They split on the nature of paternal authority. Some – like Filmer, Saravia and a number of other English absolutists – argued that in essence fatherly power is political. They concluded that the first fathers were kings, and that the first kings therefore did not draw their powers from popular consent. Others – including Jesuits and parliamentarian pamphleteers – admitted that by nature fathers rule in the family, but denied that families are states. The father, said Samuel Rutherford, was 'a second cause' of the son, and so had power over him independently of the son's consent. But this power, he claimed, did not include the right of life and death, and therefore it was not political. 'I do not believe', he concluded, 'that, as royalists say, the kingly power is essentially . . . a paternal or fatherly power'.[28]

Hobbes agreed with neither side in this controversy. He rejected two doctrines that both sides endorsed – the notion that a father's power is independent of his children's consent, and the idea that by nature it is the father and not the mother who holds dominion over the child. He also disagreed with the parliamentarian claim that fatherly power over the family is essentially distinct from a king's power over the commonwealth. Hobbes argued that if a family is large enough to defend itself from outside attack then it is a hereditary kingdom (Lev 20: 142/105; EL2:4:10; DC9:10). Such English absolutists as Buckeridge and Jackson said precisely the same thing.[29]

Hobbes' analysis of the family was intended to be coherent with his account of government by institution and conquest. He could not afford to admit the truth of patriarchalism of the Filmerian variety, for it was wholly incompatible with his system. Patriarchalists asserted that government first began by generation and that all later rulers derived their authority from the first governor by gift, succession or usurpation. To accept this theory is manifestly to reject Hobbes' views on the origins of institutive and despotic regimes. If procreation is the source of sovereignty, then Hobbes' talk about natural equality, and much else that he says, is false or irrelevant. On the other hand, if procreation leads to powers which fall short of sovereignty – as parliamentarians argued – then it is plainly possible for the head of at least one kind of human society (namely the family) to possess only limited power. But Hobbes rejected the notion that the ruler in any independent society could hold less than sovereign power. His own theory of the family neatly evaded the implications of both patriarchalist and parliamentarian views. The crucial move he makes is to treat the family as virtually identical with a kingdom by conquest.

It was widely accepted by Hobbes' contemporaries that fathers hold power over their children through procreation (or generation). Suarez deduced the powers of fathers from generation, as did Grotius who declared that 'parents gain a right over their children by generation'. He added that *both* parents have the right, but that if any controversy arises between them, preference is to be given to the father on account of the superiority of the male sex. Filmer likewise traced the power of fathers to generation, arguing that the male contributes most to the process.[30] Hobbes, by contrast, claimed that nature gives dominion over the child to the mother rather than the father. He supported the claim with two arguments. '[T]he Infant', he said, 'is first in the power of the Mother, so as she may either nourish, or expose it' and 'if she nourish it, it oweth its life to the Mother; and is therefore obliged to obey her, rather than any other; and by consequence the Dominion over it is hers' (Lev 20: 140/103; cf. EL2:4:3; DC9:3). Secondly, 'it cannot be known who is the Father, unlesse it be declared by the Mother'. A woman may be certain that she is a child's mother, but no man can be sure that he is the father. Since the identification of the father depends upon the mother's will, the 'right of Dominion over the Child' likewise depends on her will 'and is consequently hers' (Lev 20: 140/103; cf. DC9:3).

Hobbes' claim that nature grants dominion to the mother was seen by contemporaries as an instance of his love of paradox. '[W]hether out of his kindness to the Female Sex, or rather to new and uncouth Opinions', said the cleric Gabriel Towerson, Hobbes 'prefers the Mother' as the original holder of dominion over the child.[31] In fact, Hobbes was no feminist, and no critic of the patriarchalist social structure of contemporary England. Nor was shocking the orthodox his sole point in asserting the natural dominion of the mother: the idea was not wholly heterodox, for it had parallels in Roman law and in the thinking of Selden, and it was taken seriously (though rejected) by Grotius.[32] Rather, the point of the argument was to show that the power of the *father* was derived from the consent of the mother. Everyone, including Hobbes, agreed that it was ordinarily the father who came to rule in the family. But while most accepted that his rule was by divine natural law, Hobbes' argument was designed to demonstrate that it arose from consent.

In the state of nature, said Hobbes, the parents could contract with each other for dominion over the child, but if they failed to do so it remained with the mother (Lev 20: 140/103). However, if the mother herself became subject to someone, then the child would also come under the dominion of that person – 'because whoever has dominion over a person has dominion over all that belongs to that person' (DC9:5: 'qui *Dominium* in *personam* habet, *Dominium* habet in omnia quae sunt eius'; cf. Lev 20: 140/103). Hobbes mentioned three ways in which this could happen. Firstly, the mother could be taken prisoner – or in other words, conquered. Secondly, she could be subject to a sovereign in a commonwealth. Finally, she could give herself to a man, agreeing that he rule over her (DC9:5). In the state of nature, then, men can acquire dominion over women by conquest or consent. Of course, as we have seen, Hobbes reduced conquest to consent, so the father's power over the children is a consequence of the mother's consent. This was sufficient to rebut one of the central contentions of patriarchalist political thinking – that fatherly power is not based upon consent.

Hobbes had no wish to open a back door to patriarchalism by admitting that the father always and necessarily comes to hold dominion over the mother and children. Such an admission might have suggested that there is some characteristic in human nature – comparable in its universality to the principle of self-preservation – from which we can read off political and ethical conclusions, and in particular the conclusion that fatherhood confers power. He therefore argued that there have been societies in which women exercised dominion, instancing the Amazons, and noting that in a number of contemporary commonwealths the sovereign was a woman. In the state of nature, he insisted, men and women were sufficiently equal for it to be impossible (at least on some occasions) for one to gain dominion over the other without war (DC9:3; cf. Lev 20: 139/102, EL2:4:2). However, he held that it was nevertheless usually the male who did gain dominion. Once states had been set up, he argued, the sovereign ought to ensure that children be instructed to honour their parents, and be taught 'that originally the Father of every man was also his Soveraign Lord, with power over him of life and death; and that the Fathers of families, when by instituting a Common-wealth, they resigned that absolute Power, yet it was never intended, they should lose the honour due unto them for their education' (Lev 30: 235/178). So the usual pattern was for fathers to acquire dominion over their families in the state of nature.[33]

In addition to deriving the father's dominion from the mother's consent, Hobbes claimed that ultimately parental dominion springs from 'the Childs Consent, either expresse, or by other sufficient arguments declared' (Lev 20: 139/102). This was very close to Hobbes' theory of conquest. The vanquished man is in the power of the conqueror and may save his life by covenanting to obey the victor. Similarly, the child at least tacitly consents to obey its mother because she has preserved it – or if the mother is the father's subject, obedience is owed to him. Hobbes listed two ways in which sovereignty could be acquired by 'Naturall force'. One was conquest. The other occurred 'when a man maketh his children, to submit themselves, and their children to his government, as being able to destroy them if they refuse' (Lev 17: 121/88; the Latin reads 'pater' [father] for 'man': LW3:131). In both cases, sovereign authority is created in precisely the same way, namely by the consent of the person in the power of the conqueror or parent. '[E]very man', said Hobbes, 'is supposed to promise obedience, to him, in whose power it is to save, or destroy him' (Lev 20: 140/103).[34]

So far we have seen that the details of Hobbes' thinking owe much to traditional notions. Often, he makes points which are of the highest relevance to the political and ethical debates of his contemporaries, and sometimes it is abundantly clear that one of his intentions was to contribute to these discussions. His doctrine of the reciprocal link between protection and obedience, for example, was manifestly intended to shed light on the practical problem of whether Englishmen could legitimately submit themselves to the Rump by taking the Engagement. Nor was the idea of a link between protection and obedience in the least original. To what extent is it true that his account of political obligation and of the origins of government as a whole is unoriginal? In recent decades considerable controversy has arisen amongst commentators on Hobbes over whether he was, like many of his contemporaries, a conventional natural law theorist, or whether he was a revolutionary who broke the mould of old thinking and ushered in a new and distinctively modern approach to political theorising. That problem is the subject of the final section of this chapter.

4: POLITICAL OBLIGATION

Hobbes' moral and political opinions differed from those of most of his contemporaries on a number of details. They also differed in one important general respect. For Hobbes, all moral principles

Origins of Government and Nature of Political Obligation 75

derive ultimately from individual self-preservation. For others, self-preservation was indeed important, but it was only one of a number of rules, each of which generated moral conclusions. The upshot of this is that it is doubtful whether it makes sense to describe Hobbes as having any genuine moral system at all. Normally, if we say that someone has a moral obligation to do something, we mean that he should do it whether or not it is in his interest. But for Hobbes, fulfilling a moral obligation means doing what leads to self-preservation, and self-preservation is in the individual's interest. Of course, other theorists gave people an incentive to carry out their moral obligations by claiming that if they failed to do so God would punish them either in this world or the next. 'Honesty is the best policy' was the claim, and this meant that it is always prudent to act honestly since the unpleasant consequences of dishonesty will inevitably outweigh any short-term benefits that it confers. But the usual attitude was that honesty is obligatory quite independently of any advantages that it brings. We have a moral obligation to obey the laws of nature, said theorists, and we also have a strong incentive to do so, since the consequences of disobedience will be very unfavourable for us. But though self-interest may provide a *motive* for obedience, it is not the ground of our obligation to obey.

Some commentators have argued that in Hobbes' system there is a similar distinction between the obligation to obey natural law and our motives for obedience. They claim that the obligation to obey is derived from the fact that the laws of nature are the commands of God. Self-preservation, so the argument runs, furnishes us only with a motive for obedience and not with the grounds of our obligation to obey. A major implication of this thesis is that Hobbes' theory is far closer than has usually been supposed to the orthodoxy of his contemporaries. This sort of argument was advanced by A. E. Taylor in 1938, and revised and extended by Howard Warrender in 1957. Warrender claimed that the ultimate source of obligation in Hobbes is God. Hobbes, he observed, distinguishes between two forms of 'natural obligation', both of which spring from God's power. The first of these forms might be called physical obligation, since it occurs 'when liberty is taken away by corporal impediments' (DC15:7: 'ubi libertas impedimentis corporeis tollitur'). In other words, God has set down the physical rules according to which things operate. The second – and, from Warrender's point of view, crucial – type of obligation arises in people as a result of their consciousness of their own weakness and of God's irresistible power. Reason dictates that when we have no power to resist we are obliged to obey. Warrender calls this type of obligation 'moral obligation'.[35]

According to Warrender, the whole of Hobbes' theory of political obligation follows from our obligation to obey an irresistible God. Of course, we can have no such obligation if we are unaware of God's existence. So atheists – on Warrender's interpretation of Hobbes – have no obligation to obey natural law or the sovereign. All who do acknowledge God's existence, however, perceive that the laws of nature are God's laws and that obedience to them is morally obligatory. Political obligation, Warrender argued, follows from our duty to fulfil the laws of nature, including the law prescribing that we keep our covenants. The entire system of obligation is grounded, then, upon God's commands, or perhaps upon fear of the penalties for breaking these commands.[36]

Self-preservation, he continued, enters the picture only inasmuch as it gives us a *motive* for obeying natural law. According to Hobbes' psychological theory, he claimed, self-preservation is the supreme motive for people (except for salvation, which we can ignore for the present: it is discussed at length in Chapter 6, section 2 below). Since it is rational for people to act in accordance with their best interests, and since these necessarily include self-preservation, they cannot rationally act in a way which is calculated not to preserve themselves. So we can have no adequate motive for obeying natural law if it does not in fact promote our preservation. But according to Warrender, all that follows from this is that congruence with self-preservation is a 'validating condition' of obedience to natural law – in that it makes such obedience psychologically possible. It does not at all follow that self-preservation is the *ground* of the obligation to obey.[37]

Warrender added two important contentions to his discussion of self-preservation, both of which are intended to demonstrate that individual self-preservation is not at the core of Hobbes' theory of obligation. The first is that natural law in Hobbes is concerned less with the preservation of the individual than with 'the conservation of men in multitudes'. 'The laws of nature', declared Warrender, 'are not strictly rules for personal preservation'; rather 'they are rules for the preservation of men in general'. The second, and connected claim is that individual self-preservation in Hobbes is described not as a duty but as a right (though Warrender added on the basis of some of Hobbes' other statements that even if it was a duty it would oblige us only because God has commanded it). The implication of both these claims is that the laws of nature and our duty to obey them are in no sense derived from individual self-preservation. If the laws are about 'the preservation of men in general' they cannot stem from a principle concerned only with the preservation of the individual. Again, if Hobbes had intended to ground a system of duties upon individual self-preservation, he would surely have made that principle itself a duty.[38]

According to Warrender, all obligations in Hobbes arise either through covenants or through laws.[39] Since the laws of nature are not covenants, it follows that it is as laws that they oblige – and not as theorems or dictates of reason. Hobbes tells us that it is in virtue of being commanded by God that natural laws are properly laws. Warrender drew the consequence that the laws of nature oblige us only because God has commanded them. There are a number of major problems with this highly ingenious argument. Firstly, as Quentin Skinner in particular has pointed out, if Warrender's account is correct Hobbes turns into a rather conventional exponent of natural law theory, and if this is so it is very surprising that his contemporaries failed to notice the fact. For those contemporaries believed that Hobbes *was* attempting to ground obligations upon the preservation of the individual.[40]

Secondly, the textual warrant for Warrender's thesis is dubious. Hobbes discusses the laws of nature extensively in the fifteenth chapter of *Leviathan* without mentioning God until the very end. There, indeed, he does state that if we consider the theorems of reason concerning self-preservation as they are 'delivered in the word of God, that by right commandeth all things; then are they properly called Lawes' (Lev 15: 111/80). But he places no weight at all upon this point. In the thirty-first chapter Hobbes addresses a similar topic, arguing that God rules by natural law over all who acknowledge his power (Lev 31: 246/187). Atheists, he here asserts, are not God's subjects because 'they acknowledge no Word for his, nor have hope of his rewards, or fear of his threatnings' (Lev 31: 246/186). He then proceeds to outline the duties towards God which believers in him may perceive by reason alone, and without any special revelation. Atheists, who do not believe in him, have no such duties. But Hobbes nowhere states that atheists lack obligations connected with their own preservation. If he had intended to base his whole theory of obligation upon God's commands, it is odd that he gave so little prominence to the point. Indeed, Filmer drew the conclusion that Hobbes did not really believe that the law of nature was 'given by God' at all.[41]

Warrender's thesis relies upon the claim that only laws and covenants oblige in Hobbes. In both the Roman and the common laws there was a close conceptual connection between obligation and agreements entered into by individuals – contracts or deeds.[42] The same association is to be found in Hobbes, though like many of his contemporaries he also spoke of laws as creating obligations. But did Hobbes think that the laws of nature, considered merely as theorems of reason and not strictly as laws, impose obligations? The answer to this question is yes. 'The law of nature', he said, 'always and everywhere obliges' (DC3:27: 'legem naturae semper et ubique obligare'; cf. Lev 15: 110/79). There is not the slightest hint here or elsewhere in his discussion of natural law that atheists are under no obligation to obey it. Hobbes suggested that the existence of God can be proved by reason, but argued that the necessary proof was so difficult that men could easily be ignorant of it (DC14:19 Annot; see also Chapter 6, section 1 below). By contrast, he declared that no one could be ignorant of the laws of nature, for 'to leave all men unexcusable, they have been contracted into one easie sum, intelligible, even to the meanest capacity; and that is, *Do not that to another, which thou wouldest not have done to thy selfe*' (Lev 15: 109/79). Ignorance of the laws of nature cannot excuse anyone – even an atheist – from obeying them, since no one can in fact be ignorant of them. This makes no sense unless all are obliged to obey, even without knowledge of God. The primary sense in which the laws of nature oblige in Hobbes is as dictates or theorems of reason telling us how to preserve ourselves. Those who believe that the laws are also divine decrees or that God will punish us for infringing them – either here on earth by means of such natural punishments as disease for intemperance (Lev 31: 253–4/193), or by withholding eternal salvation from us in the afterlife – may have an additional incentive to obey them. But Hobbes shows little interest in such added incentives, and bases no point of importance upon them.

Warrender's contention that Hobbes' laws of nature are 'rules for the preservation of men in general' and not just for personal preservation is difficult to sustain. As we have seen, Hobbes' lists of the laws are singularly lacking in principles which provide for the propagation of the species, or which prefer the public good to private preservation. But the point is most decisively made simply by looking once again at Hobbes' definition of a law of nature.

It is, he said, 'a Precept, or generall Rule, found out by Reason, by which a man is forbidden to do, that, which is destructive of his life, or taketh away the means of preserving the same; and to omit, that, by which he thinketh it may be best preserved' (Lev 14: 91/64). Plainly, the laws of nature *are* primarily rules for personal preservation, and only secondarily principles for the preservation of others. You should aim at peace, Hobbes is saying, because it will preserve you, though it is also incidentally true that it will benefit others.

As we have seen, Warrender argued that self-preservation in Hobbes' theory is a right not a duty, and therefore cannot be the basis of a system of duties. This argument is dubious. The right of nature in Hobbes is not simply a right to preserve yourself but rather a right to exercise your own judgement about the means most conducive to this end. It makes sense to call this a right because you can agree *not* to exercise your own judgement but instead to follow someone else's – and this is just what you do agree to when you enter the commonwealth. As we have just seen, Hobbes defines the laws of nature as deductions from the principle of individual self-preservation. They are, he said, 'derived from a single dictate of reason, exhorting us to seek our own preservation and safety' (DC3:26: 'ab unico rationis, nos ad nostri conservationem & incolumitatem hortantis, dictamine derivata'). Since self-preservation is not deduced from itself, it does not qualify as a law of nature under Hobbes' definition of that term. But it is none the less the principle from which all the laws of nature – and not just the motive to obey them – are generated.

The idea that all obligation in Hobbes stems from God's will cannot be vindicated. His conception of morality is reducible to calculations of self-interest. It is sometimes asked whether Hobbes was a natural law theorist. As we have seen, there are a great many respects in which his theory coincides with those of natural law thinkers like Suarez and Grotius, and such perceptive contemporaries as Edward Gee and Sir Robert Filmer took him to be engaged in the same kind of activity as these writers. But by attempting to make moral philosophy an objective science grounded upon self-preservation, Hobbes altered the laws of nature in a vital respect, emptying them of any specifically moral as opposed to self-interested content. Provided these facts are recognised, it seems very much a matter of taste whether we style Hobbes a natural law theorist or not.

4
Hobbes on Sovereignty and Law

Sir Robert Filmer found much to criticise in Hobbes' account of how government arose. But he was far more happy with Hobbes' treatment of the powers which rulers held once it had arisen: '[w]ith no small content I read Mr Hobbes's book *De Cive*, and his *Leviathan*, about the rights of sovereignty, which no man, that I know, hath so amply and judiciously handled. I consent with him about the rights of exercising government, but I cannot agree to his means of acquiring it'. As we have seen, Filmer was an absolutist who welcomed many of Hobbes' doctrines on the power of sovereigns. Both men set down their thoughts on this subject before Civil War broke out in England, for Filmer had completed a version of his *Patriarcha* by 1632, and Hobbes finished the *Elements of Law* by 1640. In *Leviathan*, Hobbes did indeed add to his discussion a number of references to events and debates that took place in the 1640s. But the broad contours of his argument are much the same as in the *Elements*. The immediate historical context of Hobbes' central teachings on the powers of sovereigns is to be sought in the disputes on the royal prerogative which took place in England before the Long Parliament met in November 1640.[1]

John Locke traced the origins of absolutist thinking in England to the sermons of Sibthorp and Maynwaring – two preachers who defended the Forced Loan of 1627, which Hobbes himself helped to collect. Clarendon claimed that Hobbes' 'doctrine against the Liberty and propriety of the Subject' had previously been published by Sibthorp and Maynwaring, and John Whitehall asserted that Hobbes had penned his account of property 'like a dear Son of Sibthorpe and Manwaring'. Algernon Sidney likewise grouped Hobbes with Sibthorp and Maynwaring amongst the most objectionable thinkers of the age.[2] Maynwaring had been impeached by the 1628 Parliament for his sermons, though soon after the Parliament was dissolved the king pardoned him, and a few years later made him a bishop. When the Long Parliament met, Maynwaring was arrested and imprisoned in the Tower of London. According to Aubrey, this event was instrumental in leading Hobbes to seek refuge overseas: 'he told me that Bishop Manwaring (of St David's) preached his doctrine; for which, among others, he was sent prisoner to the Tower. Then thought Mr Hobbes, 'tis time now for me to shift for my selfe, and so withdrew into France, and resided at Paris'. Aubrey observed that 'there were others likewise did preach his doctrine'. Annotating this remark, Anthony Wood added the name of Robert Sibthorp (Aubrey 1:334).

Sibthorp and Maynwaring in fact drew most of their principles from earlier theorists, both English and Continental. They were not the first exponents of such ideas in England, for there were absolutists under James I and earlier. By the time that Hobbes wrote the *Elements*, there was a flourishing tradition of English absolutist thinking. It makes sense to locate Hobbes' doctrines within this tradition – and within the wider context of European theorising. As we shall see, he did indeed adopt many notions held by other absolutists. In some respects it is true that his doctrine is the same as that of Maynwaring and Sibthorp. But in others he diverged sharply from standard absolutist theory. Hobbes held that there were defects and inconsistencies in traditional thinking on the power of sovereigns. Error on this point threatened to weaken and ultimately destroy the state. So it was his task to correct traditional mistakes and thus secure the commonwealth. We shall see how he did this in what follows.

1: SOVEREIGNTY

Hobbes argued that there must be an absolute sovereign in every state (DC6:13; Lev 18). Sovereignty, he claimed, is necessarily indivisible (EL2:1:16; DC6:11 Annot; Lev 18: 127/92–3, 225/170), and the sovereign has to be either a single person or an assembly. If sovereignty is vested in just one person, the form of government is monarchy. If it is held by a council in which every citizen is entitled to vote, the form is democracy. Where the ruling assembly consists of only some of the citizens, the government is an aristocracy (DC7:1; Lev 19: 129/94). He asserted that these are the only three possible forms of government and that mixed forms cannot exist (EL2:1:16;

DC7:4 and Annot; Lev 29: 228/172). A ruler whose power is subject to limitations imposed upon him by some person or group – say by the pope, or by the two Houses of parliament – would not be sovereign; in this case sovereignty would lie with whoever imposed the limitations (DC6:18). In ancient Sparta, for example, the authority of kings was subordinate to that of ephors, who therefore possessed sovereign power (Lev 19: 135/98–99).

Hobbes argued that sovereignty was not only indivisible and unlimited, but also perpetual; that is to say, the body which held sovereignty had to continue in being. If a ruling assembly dissolved itself without transferring its powers to anyone else, the commonwealth ceased to exist and a state of war resumed (DC6:17). A ruler might delegate authority to subordinates, but could not alienate any of the powers which were essential to sovereignty without alienating sovereignty itself – and so abdicating. If the sovereign granted his subjects any liberty which was incompatible with sovereignty, they had a duty to ignore the grant and to continue to obey his commands (DC7:17; Lev 18: 127/93; 27: 209/157).

All of these doctrines were well-known in the early seventeenth century, and indeed long before then. The Frenchman Jean Bodin made much the same points in his *Six livres de la république* (Six books of the commonwealth), which was published in 1576 and frequently reprinted.[3] Bodin is sometimes seen as the first architect of the theory of sovereignty. In fact, the idea was well enough known in the Middle Ages,[4] and English thinkers were quite capable of expressing it even without drawing on Bodin – though many did in fact cite him, including the Civil War pamphleteer Sir John Spelman who borrowed from Bodin to support his royalist views, and Peter Heylin, who cited Bodin frequently.[5] But it was Heylin's friend Sir Robert Filmer who betrayed the greatest signs of dependence upon Bodin, quoting him extensively in his *Patriarcha*, and in 1648 publishing a pamphlet which consists of nothing but extracts from the *Six livres*.

Hobbes did not cite Bodin in *Leviathan* or *De Cive*. But in the *Elements* he drew on him to demonstrate that sovereignty is indivisible (EL2:8:7). The same doctrine was voiced by Hadrian Saravia in his *De imperandi authoritate* of 1593. Like Bodin, Saravia held that only the three pure forms of government were possible; mixed government could never work. He argued that a monarch whose power was subject to limitations imposed upon him by some other human authority was no true sovereign, and cited the example of the kings of Sparta. Sovereignty, he claimed, was essential to human society, for without it social relations could not last long, since people would be led into conflict with one another by love of liberty. Though liberty was commonly praised, he said, it generally amounted to nothing more than a prideful reluctance to obey others, and stemmed from self-admiration, which was the source of all evils, and which had led to Adam's fall. Like Saravia, Filmer inveighed against those 'who magnify liberty as if the height of human felicity were only to be found in it – never remembering that the desire of liberty was the cause of the fall of Adam'.[6] In Hobbes' system, likewise, it is natural love of liberty, pride, and other passions which make coercive government necessary (Lev 17: 117/85). Leviathan (the sovereign) is 'King of the Proud' (Lev 28: 221/166). The characteristically Hobbesian doctrine that any government is better than none (Lev 19: 128–9/94) was equally characteristic of Filmer, Saravia, and of many others. As the well-known divine Thomas Adams put it in 1633, 'Any King is better than no King, Tyranny is better than Anarchie'.[7]

The civil lawyer William Vaughan held that only the three pure forms of government are possible, and Hayward – another civil lawyer – took precisely the same line, asserting that 'the rights of Soveraignty . . . are nothing else, but an absolute and perpetuall power, to exercise the highest actions and affaires in some certaine state'. A state, he said, must be either a monarchy, an aristocracy or a democracy. He argued that sovereignty consists of a number of powers which are essential to good government and which the sovereign cannot alienate. Though he did not specify exactly what these powers are, others regularly did so, listing such rights as making laws, waging war, levying taxes, and coining money. James I alluded to this notion that sovereignty consists of certain inalienable powers when in 1621 he declared that the House of Commons had recently attempted to usurp all 'the highest points of Soveraigntie . . . except the striking of Coine'. They had laid the grounds, he asserted, 'for future times to invade most of those Rights and Prerogatives annexed to Our Imperiall Crowne, as bee the very markes and Characters of Monarchie and Sovereigntie'.[8]

Hobbes' claim that sovereigns possess some inalienable powers was traditional enough. His lists of these powers (EL2:1:7–19; DC6:5–18; Lev 18: 121–6/88–92) largely coincided with those of Bodin and like-minded authors. He did indeed argue that the king *could* alienate the right to coin money (Lev 18: 127/92; cf.

DC14:20), while Bodin denied this. But both men classed amongst the essential attributes of sovereignty legislation (DC6:9; Lev 18: 125/91), making war and peace (DC6:7; Lev 18: 126/92), appointing counsellors and officers of state (DC6:10; Lev 18: 126/92) and acting as the supreme judge of breaches of the law (DC6:6, 6:8; Lev 18: 125/91; 126/92).[9] The details of Hobbes' treatment of such powers sometimes betray the influence of recent English circumstance; for instance, his insistence in *Leviathan* that the militia should be under the sovereign's control patently alludes to the controversy on this point which heralded the outbreak of the Civil War (Lev 18: 126/92). But for the most part Hobbes' list of the powers was conventional.

In *Leviathan*, Hobbes claimed that the events of the Civil War period had made almost everyone aware of the need for a sovereign in each state. He predicted that when peace was restored in England the sovereign's inseparable rights would be 'generally acknowledged'. Civil war, he said, would not have broken out in the first place if most people had not erroneously believed that sovereign power was 'divided between the King, and the Lords, and the House of Commons' (Lev 18: 127/93). Quite possibly, the idea of absolute and indivisible sovereignty did become more common in the 1640s and 50s, though it was not 'generally acknowledged'. Certainly, in the years before 1642 many held that the king shared sovereign power with the two houses of parliament. In the House of Commons under James I and his son it was commonly said that the king could not make law or levy taxes except with the consent of parliament. People took this line not because they were ignorant of the Bodinian theory of sovereignty, but because they rejected it – on the grounds that mixed and limited government *was* possible, and that in England royal power was limited by fundamental laws which granted the subject certain rights that no king could infringe.

As we have seen, when Charles I was preparing for war in 1626–7 he raised money by the Forced Loan, imprisoned refusers without cause shown, billeted troops upon civilians, and used martial law in time of peace. The legality of these actions was doubtful. Seeking taxes, the king summoned parliament in 1628. It soon became clear that many members of the Commons greatly resented his recent policies. The House approved a document known as the Petition of Right, which condemned the policies as illegal. They waited until the Lords and the king assented to the Petition before voting taxes. It was clear that Charles was eager to lay hands on the taxes, but reluctant to endorse a document which declared that he had flouted the law and which purported to show that he could not revive his policies in the future. In these circumstances, the Lords suggested that a clause be added to the Petition, stipulating that the document left 'entire' the king's 'sovereign power'. The Commons rejected this on the grounds that the clause effectively subverted the Petition. Bodin argued that sovereignty 'is free from any condition'. So to grant that the king possessed sovereignty was to admit that the various provisions of the Petition did not bind him. The House of Commons rejected the clause not because they failed to grasp Bodin's theory, but because they did not believe that the king held Bodinian sovereignty.[10]

Ultimately, Charles I assented to the Petition, even without the added clause. In doing so, however, he told Parliament that he felt sure they had no wish to hurt his prerogative; moreover, he added, they could not do so even if they wanted to. In other words, the king retained his sovereign powers despite assenting to the Petition. Archbishop Laud argued likewise when he wrote on his copy of the Petition that 'Saving the right of our crown (*Salvo jure coronae nostrae*) is intended in all oaths and promises exacted from a sovereign'. A little later, Sir Francis Kynaston made the same point, claiming that 'saving the right of the crown' 'must be expressed or implied' in any royal grant – and he held that the crown's right included the power to tax and make laws without the subject's consent. Hobbes was saying nothing novel when he declared that a sovereign could not grant away powers essential to sovereignty unless he abdicated (Lev 18: 127/93).[11]

It is sometimes suggested that the notion that succession to the crown is by indefeasible hereditary right was one of the key doctrines of early modern absolutist thinking.[12] Bodin held that the natural mode of succession was primogeniture, but argued that nature excluded the female line, equating the Salic Law of France with natural law. English writers rejected Bodin's thesis that women were excluded from the succession by the law of nature – not surprisingly, since England was ruled by two queens in the later sixteenth century. But they usually accepted that primogeniture was the most natural means of succession. Some, however, argued that the king himself could appoint his successor; only if he failed to do so would primogeniture apply. This was the position of Filmer and Saravia. It was also the position of Hobbes himself. For he said that unless the king appointed his successor by will, or by tacitly approving customary rules on the question, primogeniture would come into operation (DC7:15; 9:15–17; Lev 19: 137/100–1). So there was little that was innovatory about his teaching on this question.[13]

The same goes for much of what he had to say in favour of the idea that monarchy is the best form of government – though he dismissed common analogical arguments in support of this thesis (DC10:3). Hobbes claimed that 'in all great dangers and troubles' the sovereign assembly in any large non-monarchical state needed to appoint dictators 'which are as much as Temporary Monarchs' (Lev 19: 133/97–8). So monarchy was the best form of government in time of war – and therefore always, since commonwealths (in Hobbes' view) were in a continual state of nature – or war – with one another (DC10:17). Bodin, Saravia and Filmer likewise argued in favour of monarchy by observing that democracies had appointed dictators in war-time.[14]

Again, Hobbes argued that in a monarchy 'the private interest is the same with the publique'. Where an assembly ruled, its individual members had private interests which might lead them to vote against the public good. But the 'riches, power, and honour of a Monarch arise onely from the riches, strength and reputation of his Subjects. For no King can be rich, nor glorious, nor secure; whose Subjects are either poore, or contemptible, or too weak through want, or dissention, to maintain a war against their enemies' (Lev 19: 131/96). Filmer employed the same argument, claiming that self-interest would lead a monarch to provide for the prosperity of his subjects. 'It is the multitude of people and the abundance of their riches', he said, 'which are the only strength and glory of every prince'. Charles I himself said much the same thing, declaring that no king was so great 'as he that was King of a rich and free people', and acknowledging that it was therefore in his own interests to promote his subjects' welfare. The idea was commonplace.[15]

Bodin asserted that 'the bringing up of youth' is now 'neglected', though it should be 'one of the chiefest charges of a commonweale'.[16] English writers commonly granted the monarch power to provide for the education of his subjects in true doctrine, and to prohibit the teaching of seditious ideas. Elizabeth and the first two Stuarts censored the press and encouraged the clergy to instruct their flocks in the duty of subjects towards rulers. They also intervened in the affairs of the universities, suppressing the teaching of seditious doctrines. So in arguing that the sovereign could and should control teaching and preaching (DC6:11; Lev 18: 124/91), Hobbes was saying little that was new. Others were aware that opinions influenced actions and that control of the pulpit was highly useful to anyone who wanted to control the population as a whole.[17] But Hobbes did place unusual stress upon this point. His discussion of why commonwealths decay was conducted largely in terms of 'the poyson of seditious doctrines' (Lev 29: 223/168).

Hobbes parted company with ordinary absolutist thinking in the account which he gave of what doctrines are seditious. Most absolutists held that the sovereign could never be resisted actively, but that subjects should passively obey him if he commanded them to break the laws of God or nature. Passive obedience meant disobeying, but meekly accepting the penalty. Subjects, they held, possessed rights – including rights of property – against the sovereign. True, these rights were unenforceable, in the sense that no one was empowered to coerce the sovereign into respecting them. But a ruler who infringed his subjects' rights would be acting wrongly. If the king preferred his own good to that of the public he would be a tyrant, and though his subjects could not resist him, they would be aware of his wickedness, and of the unpleasant fate which God held in store for him unless he repented and mended his ways. In monarchies where the king ruled over free subjects, he was bound to honour their rights of property and would sin if he taxed them without their own consent – except in cases of necessity. A king could sinlessly take his subjects' land or goods at will only in a despotic or herile monarchy – that is, one in which all land and goods were his property.[18]

Hobbes rejected the idea that subjects may sometimes give passive but not active obedience to the sovereign (DC14:23; Beh 49; but an apparent exception is in DC18:13). Subjects, he claimed, authorise all of the sovereign's acts when they institute the commonwealth; so the sovereign cannot do injury or injustice to the subject – since no one can injure himself (Lev 18: 124/90; 21: 148/109). Subjects, he said, have rights of property against each other but not against the king. The king, therefore, can without injustice take the subject's property whenever he wants. It follows that in all kingdoms the power of the monarch is despotic, and that the alleged distinction between despotic and non-despotic monarchy is unwarranted (Lev 20: 139/102).[19]

Not all of this was wholly novel. The Jacobean bishop William Barlow argued that Catholics could not justifiably complain of the laws passed against them in Parliament, for they had implicitly 'given their consent' to those laws, since Parliament represented the whole population including themselves. Marsilius, Hooker and others claimed that the acts of our representatives are our own acts. Filmer held that rights of property originally flowed from the will of the first king – Adam. Maynwaring argued that kings cannot commit injustice towards their subjects – an opinion which got him into trouble in the House of Lords in 1628.[20] Some thinkers reduced the subject's right of passive (dis)obedience to such an extent that there is little practical difference between their doctrine and that of Hobbes. The Laudian cleric Thomas Jackson insisted that the subject would not be warranted in disobeying the prince unless he was *certain* that what the latter commanded was unjust. Moreover, he argued that in some cases subjects were bound to obey even an unjust order – for instance if disobedience would scandalise others.[21] The royalist divine Pierre Du Moulin the younger carefully distinguished between the consequences of an action and the action itself. Subjects, he argued, should not concern themselves with consequences: 'as long as the thing commanded is lawfull in it selfe, we are not answerable of the consequence that may follow, but they that command it'.[22] So if I obey the sovereign's command to shoot in a certain direction, and it happens that someone gets killed, it is the sovereign who takes responsibility for this consequence.

Filmer argued similarly. The 'sanctifying of the sabbath is a divine law', he declared. But he added that there were circumstances in which this obligation could licitly be ignored, citing Christ's own remarks upon this point (Luke 14:5). If a master commanded his servant not to attend church on the sabbath, the servant would be bound to obey, for it might be that a relevant circumstance had arisen. The 'servant', declared Filmer, 'hath no authority or liberty to examine and judge whether his master sin or no in so commanding'. If the command was in fact sinful, it would be 'the sin of the master, and not of the servant'.[23] Since there are not many acts which are unlawful in all circumstances, there will be few cases in which subjects or servants may refuse to obey their sovereigns or masters.

All this is clearly close to Hobbes' teaching. But these writers still admitted that there are laws (of God and nature) which we should obey in preference to the sovereign's edicts – though their casuistry reduced the practical significance of the admission. Furthermore, they allowed the subject certain (albeit unenforceable) rights against their sovereigns, and in particular rights of property. In Hobbes' view, subjects hold no such rights (except the right of self-defence). His reflections on politics stemmed from his thoughts on property (DC epistle dedicatory, 9), and his doctrine of property was one of the most striking aspects of his theory.

2: THE LIBERTY AND PROPERTY OF THE SUBJECT

There was wide consensus amongst Hobbes' contemporaries that rights of property exist by a law higher than that of the state. 'Property' (or 'propriety') meant private property, and was contrasted with community (or common ownership). To assert that someone had property in something (or that a thing was someone's property) was to affirm that at least in ordinary circumstances he could dispose of it as he pleased and that it could not be taken from him without his own consent. There was general agreement that it was the law of nature which prohibited the taking of property without consent, and that this law had been confirmed by the eighth of the ten commandments – 'Thou shalt not steal'.[24]

Hobbes also held that the eighth commandment incorporated a natural law. But he argued that it is the sovereign's will alone which defines whether something is a person's property; so no one holds rights of property against the sovereign. Others claimed that there were rules separate from the sovereign's will defining what was a person's property, and they concluded that the king could not justly take his subject's goods *at will*. Some said that since the welfare of the community was to be preferred to that of the individual, the king *could* take the subject's goods without consent in what he deemed to be emergencies. But in England a number of lawyers and members of the House of Commons argued that subjects held an 'absolute property' in their lands and goods and that the king could in no circumstances take them without consent. Hobbes regarded the idea of absolute property as extremely pernicious, and counted it amongst the doctrines tending 'to the Dissolution of a Common-wealth' (Lev 29: 224/169; cf. DC12:7, EL2:5:2, 2:8:4 and 8).

It was commonly said that private property had been introduced before the institution of commonwealths, either by agreement or by the natural means of occupation. That ownership arose naturally from occupation was stated by the ancient Roman lawyer Gaius, and often repeated.[25] The most frequent line was that the earth had at first been held by mankind in common, but that private property had soon been found convenient; consequently, the earth had been divided up by the agreement of people everywhere. There were differences of emphasis on whether the introduction of private property had been necessary or merely convenient, and on the implications for property of the Fall. Aquinas held that private property is necessary to human life. William Ames declared that it was 'both just and necessary' 'by reason of the multiplication of mankind, and the prevailing of iniquity'. The Fall had made men evil, and apt to exploit common ownership to benefit themselves, while the growth of population had increased competition for the land and its fruits, and made it sensible to divide them in order to avoid conflict. The Elizabethan civil lawyer Richard Cosin made precisely the same points.[26]

Ames and Aquinas argued that private property was necessary. Selden and Suarez preferred to stress its convenience. Suarez took issue with the claim of the medieval thinkers Duns Scotus and Gabriel Biel that communal ownership had been obligatory before the Fall. He argued that before, as after, Adam's fatal sin both common and private ownership had been permitted by natural law.[27] Selden declared that natural and divine law neither commanded nor forbade private property. Adam, he said, had held property in the whole world, but after the Flood Noah and his sons became 'joint-tenants' of the earth. Later, private property was reintroduced when all agreed to renounce common ownership, and by 'some compact or covenant' concurred in recognising that occupation gives rise to property. Grotius claimed that at first the earth had been held in common, but that private property was introduced either by an express pact to divide the earth, or by a tacit pact recognising that first occupation confers property.[28]

For Suarez, Selden and Grotius, natural law did not absolutely require the introduction of private property – and this opinion was commonplace amongst Hobbes' contemporaries. Suarez noted that some earlier jurists had endorsed the opinion and deployed it to argue that monarchs *could* take their subjects' property at will. These jurists, he said, asserted that rulers are bound only by natural and positive divine laws, and concluded that these laws do not safeguard the subject's property against their ruler. According to Suarez, the argument is feeble. For natural law *does* forbid theft. The law of nature may not, indeed, prescribe private property, but once it has been instituted natural law condemns everyone (including a ruler) who arbitrarily seizes the property of another. This position was wholly conventional.[29]

In late-medieval and early-modern Europe the usual doctrine was that the earth and its fruits had initially been held in common, and that they had later been divided by mutual agreement. But some writers maintained that the notion of original community was mistaken and that private property had prevailed from the first. This was the opinion of the anonymous pamphleteer who called himself Eutactus Philodemius, and who took issue with Hobbes' *De Cive* on the question of original community. Eutactus rejected the idea that property 'is not from nature but from the pact and consent of man who is forced thereto by a kind of necessity for prevention of those evils, which would necessarily be the consequents of having all things common'. He used the eighth commandment to show that private property was natural. Bodin had similarly affirmed that it would be 'against the law of God and nature' for all things to be held in common, arguing that the decalogue condemns theft, and that theft presupposes private property.[30]

The same claim was used by a number of the Levellers during the famous Putney debates of 1647, when Leveller plans for constitutional reform were discussed. When some of their critics suggested that these schemes undermined property, Levellers countered by claiming that human constitutions were founded upon consent, and could be altered if superior arrangements became available; but property had its basis in divine law, as the eighth commandment indicated, and nature gave rights of property to the possessor. At Putney, the Levellers' arch-opponent Henry Ireton responded by voicing the commonplace notion that 'property is of human constitution'. He concluded that if political constitutions could be overturned then so could property – for both were based on consent. By grounding private property in the law of nature and in natural means of appropriation the Levellers tried to meet this objection.[31]

Earlier, Richard Cosin similarly cast doubt on the idea of original community by asserting that from the beginning 'none without injurie and breach of the lawe of nature, might invade the possession of that, which another had first taken up for his owne use, so long as he would imploie it'.[32] Locke later worked out a detailed theory of natural appropriation. One merit of the notion was that it avoided the need to suppose that the entire population of the world had at some point in the past met together and unanimously consented to the introduction of private property – a rather implausible supposition, as Filmer gleefully noted.[33] But without such an instantaneous and universal act it was difficult to see how consent could introduce private property – since all those who did not consent would still retain their rights to hold everything in common.

Though there was dispute on the details, it was generally agreed that private property at first came into existence either by natural appropriation or by consent – which was itself underpinned by the natural law that people should abide by their contracts. In other words, some people said that natural law set up rules on how property could be acquired (for example specifying that rights of property arose from occupation), while others argued that the rules were instituted not by nature but by consent, and that occupation conferred property because people had consented to this arrangement. In either case, the way in which the rules were instituted was quite independent of how governments were set up – and many said that private property in fact ante-dated the institution of governments. Indeed, the protection of rights of property was commonly regarded as a major function of government – and sometimes as its principal purpose. The natural law prohibiting theft outlawed the arbitrary taking of a subject's property even by the sovereign. So except in herile or despotic monarchies – where the king was the sole proprietor – subjects held rights of property even against their monarch. Saravia typically observed that this applied even in monarchies which had at one time been herile but where the sovereign had later granted his subjects rights of property, for princes are bound to honour their agreements. It did not necessarily follow that subjects could enforce their rights, either by violence or by judicial means. Bodin and Saravia both insisted that property rights applied against the king, but that he could never be actively resisted even if he infringed them. Nor could subjects take the king to court except with his permission. But a king who wantonly seized his subjects' property would none the less be guilty of injustice towards them.[34] Of course, Bodin and Saravia were absolutists. Others sometimes argued that a king who infringed rights of property *could* be resisted. The notions that Charles I – or at least his evil advisers – planned to undermine the property rights of English subjects and that it was therefore legitimate to fight a defensive war against him – or them – were amongst the main planks of parliamentarian ideology in the Civil War.

Since property was founded on a law higher than that of the sovereign, people argued, he could not justly take it at will. But this did not mean that rights of property were wholly disconnected from human lawmaking or from the powers of rulers. Firstly, sovereigns protected their subjects' property against domestic and foreign depredations, and it was reasonable that they be recompensed for this. Moreover, in times of emergency it might turn out that the sovereign would be unable to defend either the goods and land of his subjects or even their lives unless he were empowered to tax them without their consent.

Secondly, though property could exist outside the commonwealth, and could be transferred by gift or covenant, disagreements might easily arise over whether something was a particular person's property, over whether he had in fact covenanted to part with it, and so on. Indeed, it was acknowledged that one of the main functions of the civil magistrate was to resolve disputes on such points by defining just what was to count as valid acquisition or transference of property. Of course, human law could not abrogate natural laws relating to property, but it could add to them more detailed rules which promoted the public good either by bringing precision to transactions involving property – and thus reducing the likelihood of disagreements – or by restricting the acquisition or transfer of certain commodities. For instance, Aquinas argued that sovereigns could limit the sale of land to ensure that depopulation did not occur,[35] and early modern sovereigns frequently forbade merchants from exporting gold bullion so that their realms would not suffer dearth of this precious metal.

The idea that rulers should be reimbursed for protecting their subjects was widely accepted. Absolutists commonly argued that kings could levy impositions upon exports and imports, and that in cases of necessity they were empowered to tax their subjects even without consent. It was, they said, the king alone who was to judge what constituted necessity.[36] Bodin insisted that the sovereign ought ordinarily to obtain the consent of his subjects before he taxed them, and the same point of view remained commonplace in early seventeenth-century England – though in France the idea that consent is normally required rapidly withered away.[37] English absolutists typically claimed that sovereigns possess the power to tax without consent in what they take to be cases of necessity, arguing that the state would be fatally weakened if kings had no right of emergency taxation. This was the line taken by

Maynwaring, Sibthorp, and Charles I himself. Suarez and Grotius similarly claimed that the power to tax without consent in cases of necessity was a mark of sovereignty – but their views on the origins of government left open the possibility that the people had reserved this power from their rulers.[38]

The first two Stuart kings not only asserted their right to tax without consent in times of necessity, but also put the right into practice – and in circumstances where the necessity seemed far from clear to many of their subjects. In response to royal claims, lawyers and members of the House of Commons argued that subjects held what Nicholas Fuller (and Hobbes) called 'absolute property' in their land and goods.[39] There were, they asserted, no cases in which the king could take goods without consent. If he could tax in what he considered to be emergencies, then he could permanently employ the pretext of necessity to deprive his subjects of all they possessed. So if royal claims were accepted, the property rights of subjects would be destroyed, turning free men into villeins (whose goods could be arbitrarily seized), and making England a despotic monarchy. In the 1640s, some authors argued that villeinage and despotic government were contrary to nature.[40] Earlier, there was wide agreement that in England at least the monarch did not hold despotic power.

The point of Hobbes' account of property was to refute both the parliamentarian theory of absolute property and the conventional absolutist idea that taxation without consent was unjust except in cases of necessity. Like Filmer, he wanted to show that individuals had not possessed property rights before the institution of commonwealths. Filmer claimed that political power had existed from the very beginning, and that Adam had been both the proprietor and sovereign of the world; so all later rights of property were necessarily derived from Adam, the first sovereign.[41] Hobbes rejected patriarchalist thinking of the Filmerian variety, but reached some similar conclusions from very different premises. He began his analysis by accepting a doctrine which Filmer vigorously rejected but which was widely endorsed – namely that nature gave all things to everyone in common (EL1:14:10; DC1:10; 'natura dedit omnia omnibus').[42] His next step was to argue that until the institution of sovereigns who would coerce people into abiding by their covenants it was impossible to introduce property.

Others commonly said that the erection of governments made property more secure, but that binding covenants – and therefore property – were possible outside the state. By contrast, Hobbes argued that the distribution of property was wholly a consequence of the sovereign's will. Commonwealths, he asserted, were instituted when people transferred their rights to a sovereign. Thereafter the sovereign alone possessed the right of nature, which was a right to all things. So subjects could hold no property against him, and the rights which they held against one another were necessarily defined by his will, expressed in the laws. Others argued that human laws regulated the acquisition and transference of property, but that the institution itself was underpinned by natural law and that anyone – including a sovereign – who infringed property rights would be guilty of injustice.[43] For Hobbes, a subject who illegally took another's property was indeed guilty of breaking natural law – for this law requires us to obey our rulers; but the sovereign could never violate rights of property. He might deal inequitably with his subjects – say by showing favouritism in how he distributed taxes – and so break a different law of nature. He could never injure them, however, since they authorised all his actions.

A main purpose of Hobbes' teaching on covenants and on the origins of government was to demonstrate that no one could have rights of property against the sovereign. Subjects might, indeed, defend themselves against attack by the sovereign, and except in cases of necessity they might refuse to perform dangerous and dishonourable offices. But all other liberties of the subject were derived from the law – since liberty was but the silence of the law (Lev 21: 150–3/111–13). It was the law, too, which granted people property. To show that the subject's liberty and property were compatible with the unlimited power of the sovereign, it was therefore necessary to demonstrate that the law was nothing other than the sovereign's will. In Hobbes' view, to say that something is a man's property is simply to say that the sovereign has declared that he will protect the man's exclusive use of the thing against incursions by others. Similarly, to pronounce that someone has a liberty to do something is to assert that the sovereign has not in fact forbidden the action in question. But English lawyers and parliamentarians commonly argued that the law was not just the will of the prince. Rather, it was a set of ancient and rational customs which the king cannot abrogate and which circumscribe his powers. The law, they said, grants English subjects liberties and property which the king can neither justly nor validly infringe. Much of Hobbes' discussion of law was aimed at refuting this thesis.

3: LAW AND THE SOVEREIGN

To demonstrate that the subject's legal liberties are derived from the sovereign's will, Hobbes argued that law simply is the command of the sovereign. This was a commonplace notion amongst absolutists, and a doctrine long familiar in the scholastic and civilian traditions. 'What the prince decrees has the force of law', said Justinian's *Institutes*. According to Bodin, law means 'the commaundement of him which hath the soveraigntie'. James I and Saravia took the same line, spelling out the implication that in England parliament's role in legislation was purely advisory.[44] Filmer and Kynaston stressed this latter point, and Kynaston also gave reasons for thinking that an assembly such as parliament was not a particularly efficient body at giving advice.[45] Hobbes argued similarly, and like Kynaston he lost few opportunities to criticise the ideas of Sir Edward Coke – whose theories were the main target of attack in the exposition of the nature of law which Hobbes put forward in *Leviathan* and in the *Dialogue* on the common laws.[46]

The idea that law is the command of a lawgiver was conventional amongst absolutists, and it also featured in the thinking of many theorists who rejected the Bodinian theory of sovereignty. In his highly influential treatise *De legibus* (On laws), Suarez held that rulers generally derive their powers from the grant of the commonwealth, and that the extent of these powers depends upon the terms of the grant. Grotius argued that sovereignty could be divided between king and people – for instance if the people reserved some powers to themselves and gave others to their ruler. Yet both Suarez and Grotius maintained that human law was the will of the legislator. Grotius contrasted command with counsel, insisting that human law is command, and that it springs from the will of a legislator who has the power to enforce his orders. A king's laws are one thing, he asserted, and his contracts another.[47] Hobbes likewise contrasted law with contract (DC14:2) and command with counsel (DC6:19; 14:1). He claimed that human law requires promulgation (Lev 26: 187–8/140), that it must prohibit (or command) actions performed after and not before such promulgation (Lev 27: 203–4/153), and that it must be backed up by coercive sanctions (Lev 27: 203/152). Suarez, who self-consciously attempted to summarise scholastic tradition in *De legibus*, adopted all these positions. He declared that the consent of subjects is not always necessary for lawmaking, since 'the will of the prince is sufficient' to legislate.[48] Hobbes argued similarly, drawing the consequence for English circumstances that parliament makes laws only if it is sovereign (Lev 26: 186/139).

So, much of what Hobbes had to say about law was said also by other absolutists and by such influential thinkers as Suarez and Grotius. Of course, there were important points on which he diverged from the views of all these writers. Others held that natural law consisted of moral absolutes which had to be obeyed even against the command of the sovereign, while Hobbes argued that this was not so, for 'the principall' law of nature was 'that we should not violate our Faith, that is, a commandement to obey our Civill Soveraigns, which wee constituted over us, by mutuall pact one with another' (Lev 43: 404/322). Natural law prescribes absolute obedience to the sovereign except when he commands us to perform self-destructive acts.

Again, Suarez, Bodin and the rest held that we have a duty to obey divine positive laws – supernaturally revealed by God, and applying to a particular time and place – even if this means disobeying the sovereign. Hobbes rejected this argument on the grounds that all laws require sufficient promulgation but that we can never be sure whether a private person who claims to have had a supernatural revelation is speaking the truth or not, so that there can be no sufficient promulgation in this case. On the other hand, it may happen that the sovereign gives us orders and tells us that God has revealed them to him. In this event, subjects would be bound to obey the commands because they spring from the sovereign's will but would not be obliged to believe that God had revealed them – for no one can be obliged to believe anything. Divine positive law therefore collapses into human law (Lev 26: 197–8/148–9; for further relevant material on God, the sovereign, and belief, see Chapter 5, section 1, and Chapter 6, section 2 below).

Hobbes dissented from virtually everyone in denying that either divine positive or natural laws can sometimes justify at least passive disobedience to the sovereign – though as we have seen the practical implications of this should not be exaggerated. He dissented from thinkers like Grotius and Suarez – and agreed with men such as Bodin and Filmer – in holding that sovereignty is indivisible and concluding that the sovereign's lawmaking power cannot be circumscribed by contractual limitations imposed upon him by his subjects. To support this viewpoint he repeatedly cited the biblical maxims that no one can serve two masters (Matthew 6:24; Luke 6:13; DC9:1; Lev 29: 227/171) and that a kingdom divided cannot stand (Matthew 12:25; Lev 18: 127/93; 29: 227/171). By equating law with the sovereign's will and insisting that sovereignty cannot be divided, he was able to reach the conclusion that the law in England granted subjects no inviolable liberties against the prince. This struck at one of the central tenets of the parliamentarian opposition to Charles I.

To vindicate the thesis that the subject holds inviolable liberties, many argued that the sovereign's power is limited by contract. Others reached the same conclusion by denying that human law is merely the will of a lawgiver. This was the line taken by Sir Edward Coke and a number of other early seventeenth-century lawyers. They asserted that the English common law was superior to any merely man-made legislation since it was immemorial custom which by surviving for so many centuries had proved its unique suitability to the people of England. The common law encapsulated the quintessential wisdom of long ages, they said, and was therefore to be preferred to the decrees of any modern man or assembly, however wise. Though parliament might remedy particular defects in the common law, they argued, the fundamental principles of that law were superior to statute. It was, indeed, from the common law that parliament derived its powers.

Coke and others affirmed that the common law was the law of reason. They said that any custom which conflicted with reason could not be law. The kind of reason which they claimed was needed to understand the law, however, was not the ordinary, every-day reason that all people possessed, but the 'artificial reason' of lawyers who had spent many years in legal study. So the king, who was not a lawyer, could not himself interpret the laws but had to commit this task to qualified men. The laws, said lawyers, defined the powers of the king and the liberties of the subject. In particular, they forbade the king from legislating or taxing his subjects without their consent in parliament. Writing against Hobbes' *Dialogue*, the great lawyer Sir Matthew Hale reasserted much of this theory – though unlike Coke he incorporated contractualist elements into it.[49]

Hobbes rejected the claim that custom by itself could have the force of law: 'I deny that any Custome of its own Nature can amount to the Authority of a Law' (Dialogue 96). In places where customs became law, this happened not 'from Length of Time' 'but by the Constitutions of their present Sovereigns' (Lev 26: 186/139). He noted that English lawyers 'account no Customes Law, but such as are reasonable'. Hobbes accepted this contention but argued that it is the task of the sovereign and not of lawyers to determine what is reasonable (Lev 26: 184–5/138). Coke's notion that individual lawyers should interpret the law according to the artificial reason of each was quite unworkable, 'for then there would be as much contradiction in the Lawes, as there is in the Schooles'. Lawyers would inevitably disagree with each other and this would lead to anarchy (Lev 26: 187/140). Coke, he said, was right to call the common law the law of reason – but only 'when by it is meant the Kings Reason' (Dialogue 143).[50]

Hobbes drew on Justinian's *Institutes* to show that the Roman law was divided into seven varieties, and argued that the same division was also largely applicable to English circumstances – a contention ill-calculated to please common lawyers, who often boasted that their laws were far more excellent than civil law, and very different from it. The Romans, he said, recognised that custom had the force of law only if it was endorsed by 'the tacite consent of the Emperour', and indeed they derived all human law from the prince's will. Their laws, he affirmed, included edicts made by the emperor alone, and he equated these with 'the Proclamations of the Kings of England' – challenging the common English notion that kings cannot make law by proclamation (Lev 26: 196/147).[51] Like other laws, then, custom drew its force from the sovereign's will; in the case of custom, he expressed this will by silence. Custom, Hobbes concluded, 'is no longer Law, then the Soveraign shall be silent therein' (Lev 26: 184/138; cf. EL2:10:10; DC14:15).

This attitude towards custom was a straightforward consequence of the notion that law is the sovereign's command – and a consequence which was regularly spelled out by Hobbes' predecessors. Custom, said Bodin, 'hath no force but by sufferance, and so long as it pleaseth the soveraigne prince'. According to Suarez, custom requires 'at least the tacit licence of the prince' to become law. The same point was applied to English circumstances in a *critique* of the ideas of Sir Edward Coke which the Jesuit Parsons published in 1606. Parsons argued that all laws – including customary laws – 'must needs be made or admitted by some Prince or people'. The Jacobean Catholic priest Matthew Kellison claimed that custom becomes law only if the ruler does not disapprove of it – precisely Hobbes' doctrine. Filmer viewed matters similarly. 'Customs', he said, 'at first became lawful only by some superior power which did either command or consent unto their beginning'. This superior power was the sovereign, whose will alone turned custom into law.[52]

So, much of Hobbes' theory of law was far from original. True, he deviated from the thinking of writers like Bodin and Filmer in arguing that subjects may neither actively resist nor passively disobey their sovereigns (except in self-defence); for they said that there are divine laws which we must observe even if this means disobeying the sovereign, while in Hobbes' view, divine law does not warrant disobedience. He is, in consequence, often seen as a far more wholehearted advocate of totally unlimited state power than his predecessors. But though he did indeed drastically reduce the number of cases in which subjects may disobey their sovereigns, he none the less claimed that the sovereign *does* have obligations. These are the subject of the next and final section of this chapter.

4: THE DUTIES OF THE SOVEREIGN

Hobbes held that the office of the sovereign 'consisteth in the end, for which he was trusted with the Soveraign Power, namely the procuration of *the safety of the people*; to which he is obliged by the Law of Nature, and to render an account thereof to God, the Author of that Law, and to none but him' (Lev 30: 231/175). So sovereigns are bound to abide by the laws of nature, and to use their coercive powers in order to promote the welfare of their subjects. The sovereign could, indeed, never act unjustly. But he was subject to 'that principall Law of Nature called *Equity*' (Lev 26: 195/146; 30: 237/180) – and to the other natural laws.

Specific consequences flowed from this, some of which Hobbes spelled out. Sovereigns should impose taxes upon their subjects equitably, he said, and he suggested that the best means of doing this would be by taxing not what people earned but what they consumed (Lev 30: 238–9/181). In the early-seventeenth century the Dutch imposed heavy taxes on consumption, as did the English parliament during the Civil War. Hobbes argued that one advantage of a sales tax was that people would pay it relatively cheerfully, whereas in the case of other taxes they were likely to feel that 'they are over-rated, and that their neighbours whom they envy, do thereupon insult over them' (EL2:9:5). In England during the 1630s a great many people did indeed complain that they had been over-rated for Ship Money.[53]

Hobbes argued that '[a]ll Punishments of Innocent subjects, be they great or little, are against the Law of Nature' (Lev 28: 219/165). In his opinion, the point of punishment was to promote the public good, and this could not be done by harming the innocent. The punishment of criminals served the purpose of ensuring that the law would be better kept in the future. It was not vindictive or retributive. The same broadly utilitarian theory of punishment was common amongst Hobbes' Catholic and Anglican contemporaries, though some puritans took a different line, arguing that people had a duty to enforce the criminal laws of the Old Testament – known as the 'judicial law of Moses'. This was the opinion of the Elizabethan Presbyterian Thomas Cartwright, the New England Independent John Cotton, and others. Since God Himself was responsible for instituting the judicial law, they claimed, it was man's duty to enforce it.[54] In the 1640s some argued that if the magistrate failed to mete out the correct punishments to criminals it was up to others to do so. Since failure to do the Lord's bidding – say by not killing a blasphemer or murderer – would bring God's wrath to bear upon the people, it was in the common interest that private individuals execute God's law if the public authorities failed to do so. This position was supported by the Old Testament example of Phineas, who had employed the *jus zelotarum* (the right of the zealots) to kill some idolaters when the magistrate neglected to execute them. The argument was used to justify Pride's Purge and the subsequent execution of Charles I.[55]

So some held that punishment should take place according to a rigid, divinely established code, and without regard to changing circumstance. Hobbes rejected this position, claiming that divine positive laws were binding only if the sovereign happened to institute them (Lev 26: 197–9/148–9). The judicial law of Moses therefore imposed no obligations in modern states unless their sovereigns chose to introduce it. Most of Hobbes' Anglican and Catholic contemporaries held that the judicial law was now abrogated and argued that the right to punish was in the hands of the sovereign, who should use it to promote the public good, principally by discouraging crime. Like Hobbes, they argued that this could not be done by inflicting penalties upon the innocent. Nor, they said, did the right of punishing devolve upon private individuals if the magistrate failed to exercise it. '[A]ll punishments in a prudent government', said Jeremy Taylor, 'punish the offender to prevent a future crime, and so it proves more medicinall than vindictive'. 'The punishment of wrongdoers', said Suarez, 'is ordained for the common good of the commonwealth, and therefore is not committed to anyone except him to whom the power of the commonwealth is committed'. But though Catholic theorists commonly adopted a broadly utilitarian theory of punishment, they denied that it could ever be licit to punish the innocent. Like Suarez, the leading Anglican Henry Hammond rejected the notion that authority to punish could devolve upon private individuals if the magistrate failed to exercise it, arguing that the ancient Jewish *jus zelotarum* had been a special, extraordinary power established by God himself and that it furnished no precedent for modern states.[56] Hobbes similarly inveighed against the 'dangerous opinion, that any man may kill another, in some cases, by a Right of Zeal', and he claimed that the actions of Phineas became lawful only when they were subsequently ratified by Moses – the sovereign (Lev 'Review and Conclusion', 487–8/392–3).

Hobbes' views on punishment were not especially original. The same goes for most of what he had to say about the reasons for the decay of states, and the duties of sovereigns – which, he held, consist largely in preventing such decay (DC, chapters 12–13; Lev 29–30: 221–44/167–86). Much of his account of these reads like a commentary on recent English history – explaining why Charles I encountered such grave political problems by 1640, and how he could have avoided them. Many of the points he made were commonplace enough. He warned of the dangers that stemmed from 'the immoderate greatnesse of a Town' (Lev 29: 230/174). James and Charles repeatedly tried to limit the expansion of London. He argued that rulers should avoid making unnecessary laws (Lev 30: 239–40/182; DC13:15), and insisted that inequalities in social status were wholly dependent upon the sovereign's will (Lev 10: 65/43–4). Both positions had earlier been forcefully expressed by many, including 'our most wise King *James*' (as Hobbes called him: Lev 20: 138/101).[57] He wanted sovereigns to make provision for the instruction of subjects in their political duties (Lev 30: 234–5/178). The Tudor and Stuart monarchs made such provision – though not perhaps to the extent that Hobbes desired.

In essentials, Hobbes' opinions on the duties of sovereigns coincided with those of earlier absolutists. He argued that the sovereign ought to advance the welfare of his subjects – and specified that the state should intervene in economic life to encourage industry and agriculture, and that it should enforce uniformity in religious rites.

The idea that he was an advocate of liberalism, or a prophet of the minimalist state, is only partially correct.[58] The laws which require the sovereign to promote welfare are the laws of nature, to which he (like everyone else) is subject (Lev 30: 237/180). Of course, in Hobbes' system the laws of nature are deductions from the principle of self-preservation. It is, therefore, in order to preserve himself that the sovereign should abide by natural law. Hobbes employed two key notions at this point in the argument. The first was that 'the good of the Soveraign and People, cannot be separated' (Lev 30: 240/182). As we saw, this was a common contention. Hobbes used it to argue that it is in the sovereign's own interests to enhance the prosperity of his subjects. Secondly, he claimed that sovereigns who failed to do their duty incurred the risk of rebellion (this is a central theme of Lev chapter 30; in DC12:9 and EL2:8:2 he warns against too high taxation, since 'great exactions' can cause 'great seditions'). Of course, subjects had no right to rebel against a sovereign even if he did rule them in ways that broke the laws of nature – for example by ignoring their interests. But Hobbes acknowledged that they were likely to do so. James I had said much the same thing. Kings who broke good laws and ruled in their own interests rather than in those of their subjects, he remarked, were likely to be overthrown by those subjects, 'although that rebellion be ever unlawfull on their part'.[59]

In Hobbes' theory, the laws of nature provide a demonstrative moral code against which the conduct of sovereigns (like everyone else) may be measured. Of course, Hobbes held that we must obey our ruler even if what he commands is contrary to the law of nature (Lev 16: 113/81). Within civil society, he insisted, the sovereign is the ultimate interpreter of what is publicly to count as natural law (Lev 26: 190–1/143). But he argued that if a case arose in which the laws of the land did 'not fully authorise a reasonable Sentence' the judge was to interpret the law equitably. In other words, subordinate judges in law cases were not simply to enforce the letter of the sovereign's laws but were to construe them so that they accorded with equity, if necessary applying to the sovereign for special authority to do this (Lev 26: 194/145).

Like James I, Filmer, Bodin and others, Hobbes held that there are universal moral rules which all can perceive and which may therefore be employed by subjects to examine whether their sovereigns' laws and actions are contrary to equity. All these authors argued that the sovereign's will is none the less the ultimate criterion of what is to be enforced by the courts of law.[60] All admitted that the sovereign could in fact issue commands which contravened natural law, and that if he did so frequently he would be likely to irritate his subjects, who might rise in rebellion against him. All argued that such rebellion would be unjustified. Hobbes differed from the rest in two principal ways. Firstly, they said that subjects should not obey the sovereign's commands if they knew that this would mean breaking natural law. Hobbes argued that natural law itself requires us to obey all of the sovereign's orders unless obedience would be both dangerous (or dishonourable) and also unnecessary to the security of the state (Lev 21: 151–2/112). For instance a son could refuse to kill his father, since it would be shameful for him to do so, and the sovereign could easily get someone else to perform the act (DC6:13). As we have seen, many authors in fact imposed such large restrictions upon the subject's right of passive disobedience that in practical terms their position is not far from Hobbes'.

The second, and major innovation in Hobbes' discussion is his insistence that sovereigns can never commit injustice towards their subjects, though they may break natural law by acting inequitably: 'they that have Soveraigne power, may commit Iniquity; but not Injustice, or Injury in the proper signification' (Lev 18: 124/90). Maynwaring had made much the same claim in 1627, in order to vindicate the Forced Loan. In Hobbes' system the contention served much the same function. Sovereigns might indeed act wrongly by imposing oppressive and inequitable taxes; but taxes could not be *unjust* since the subject's rights of property and other liberties stemmed from the king's will. So those who opposed Charles I's levies – such as Ship Money – on the grounds that they infringed liberties had simply failed to understand what liberty and property are. In particular, they had failed to grasp that covenants cannot bind in the state of nature and that property cannot be instituted where there is no sovereign. The whole argument served to vindicate the financial policies which Charles I pursued before 1640. But after that year the king found it expedient to abandon those policies, and royalists distanced themselves from the kind of position which Hobbes adopted. The central message of Clarendon's critique of *Leviathan* was that its account of royal rights over property was utterly mistaken.[61] A second reason why royalists came to take issue with Hobbes' views was because of the opinions which he expressed on religion and on church-state relations. Those opinions are the subject of the next two chapters.

5
Hobbes on Church and State

De Cive is a dry, closely-argued treatise. *Leviathan* is a far more impassioned work. That is especially true of the last two parts, which discuss religion and church-state relations. After 1640 the collapse of the church courts and of censorship (which had been in the hands of the clergy) made it easy for people to publish anti-clerical and heterodox doctrines. For a while, it seemed that a harsh and intolerant Presbyterian system of church government might be established in England. But a coalition between congregationalists or Independents (supported by Oliver Cromwell), Erastians (including John Selden), and radical religious sects (such as the Baptists) ensured the defeat of the Presbyterians by 1648. In the next few years a great many authors put forward schemes for the reform of the church and universities. Hobbes was one of a large number of writers who attacked the clergy's pretensions to power, and who advocated educational reform. He hoped that the universities would abandon scholastic theology in favour of scientific research. The same hope informed the thinking of Sir Francis Bacon and of Samuel Hartlib's circle of friends, which was much influenced by Bacon.[1]

The third part of *Leviathan* is concerned especially with ideas on church-state relations. Hobbes puts forward his own theory on the topic, and refutes alternative views – particularly those of the Roman Catholic Cardinal Bellarmine. Hobbes was a materialist; he held that the notion of immaterial spirits was nonsense employed by clerics to frighten the laity into parting with power and wealth (Lev 46: 464–7/372–4). There are no incorporeal spirits, he said, and the clergy has no spiritual power – for all government is temporal (Lev 39: 321–2/248). To support his claims, Hobbes argued from both reason and Scripture. It was commonly recognised that reason and nature are insufficient to tell Christians their religious duties, or to inform them of the powers and functions of the church. Information on these matters was to be gleaned not from reason – which all men possessed – but from revelation – which God had given only to Christians, and which was inscribed in the bible. So in the third part of *Leviathan* Hobbes undertook to expound 'the Nature and Rights of a Christian Common-wealth' by drawing not only on 'the Naturall Word of God' (which was reason), but also on 'the Propheticall' word – that is to say, Scripture (Lev 32: 255/195).

Like most of his contemporaries, Hobbes claimed that reason and revelation are in harmony. There were, he admitted, 'things in Gods Word above Reason'; but there was 'nothing contrary to it' (Lev 32: 256/195). Scripture included references to a number of 'mysteries' which were 'not comprehensible', and which there was therefore no point in attempting to understand by rational means. In these cases, said Hobbes, we should refrain from contradicting the bible's words, but not labour to sift 'out a Philosophicall truth by Logick . . . For it is with the mysteries of our Religion, as with wholsome pills for the sick, which swallowed whole, have the vertue to cure; but chewed, are for the most part cast up again without effect' (Lev 32: 256/195; cf. DC18:4). Given that reason and revelation are in harmony, any principles which are demonstrable by reason remain true after people receive revelation. So all the doctrines proved in the first two parts of *Leviathan* are necessarily compatible with Scripture. Why, then, did Hobbes bother to draw on Scripture at all? And why should we bother to read the last two parts of *Leviathan*?

One point that needs to be stressed here is that Hobbes was not at all unusual in claiming that political rights and duties are derived from the law of nature, which is reason, and that grace or revelation only confirms (or adds to) but never contradicts the truths which reason teaches. The standard Catholic line was that dominion is founded not in grace or revelation, but in nature or reason. Pagan rulers were held to exercise legitimate dominion over their subjects, though they had never heard the Scriptural message. Lawful rulers did not lose their authority simply by being bad Christians – or no Christians at all. Protestants certainly placed great weight upon the bible in answering religious questions, but with few exceptions they likewise argued that dominion is founded in nature and not grace. Oliver Cromwell himself endorsed this position, though he was a religious zealot, and was much concerned with the welfare of the godly. Mere godliness, he argued, gives people no special right to rule.[2]

Some thinkers (including the influential puritan William Ames) claimed that reason is so corrupt that its precepts are reliable only when confirmed by Scripture.[3] Others – including Grotius and Chillingworth – placed greater confidence in reason. But virtually everyone held that the bible could profitably be used to confirm the truths of natural law, and also to provide information about the Christian religion – information which was compatible with reason, but which could not be deduced from it. Hobbes used Scripture in the same ways. Even if his own political doctrines were not 'Principles of Reason', he declared, 'I am sure they are Principles from Authority of Scripture' (Lev 30: 233/176). As we shall see, it is arguable that Hobbes did not in fact believe that the bible is a divinely inspired text. So why did he bother to analyse its meaning at such great length? One answer is that he intended to persuade his contemporaries of the truth of his ideas, and by so doing prevent future civil wars. Since these contemporaries regarded Scripture as authoritative, it made sense for him to discuss its meaning. Secondly, it is highly likely that he enjoyed provoking the orthodox with his idiosyncratic interpretations of the bible.

This chapter looks at Hobbes' views on church-state relations. Some of his religious opinions – on the Trinity, for instance – were strikingly original, but a large proportion of what he had to say was commonplace amongst Anglicans, and his discussion is steeped in allusions to contemporary debate. As we shall see in the second section of this chapter, Hobbes accepted much of the standard Anglican case against the views of such Roman Catholics as Cardinal Bellarmine. What he did was to develop and extend that case in order to undermine the position of the Anglicans themselves, and to deprive the clergy of all powers except those that the sovereign happened to grant them. That is the theme of the second and third sections below.

It is sometimes suggested that Hobbes not only expanded but also drastically altered his account of church-state relations in the years between writing *De Cive* and *Leviathan*.[4] In the third section we shall see that the two books put forward much the same doctrine, though it is expressed far more circumspectly in *De Cive* than in *Leviathan*. Hobbes is conventionally seen as an Erastian on questions of church-state relations. In the fourth section his thinking is compared with that of other Erastians, including

Thomas Erastus himself. As we shall see, it is fair enough to call Hobbes an Erastian in a broad sense, but he was certainly no strict follower of Erastus. His central ideas are closer to those of the great medieval thinker Marsilius of Padua than to the teachings of many seventeenth-century Erastians. But before delving into the question of how Hobbes interpreted Scripture it is worth briefly investigating the important prior problem of what he thought Scripture is.

1: WHAT IS SCRIPTURE?

Catholics observed that we cannot discover which books are Scriptural simply by looking at the evidence on this question presented in the bible – since until we know which are the Scriptural books we evidently do not know what the bible is. Some source of information outside the bible is therefore necessary in order to tell us of what books the bible consists. For Catholics, this source of information was the church and its traditions. The church, they said, was an infallible guide on what constituted Scripture and on how Scripture should be interpreted. Replying to William Chillingworth in 1638, the Jesuit Matthew Wilson argued that 'we cannot know Scripture to be the word of God, by Scripture it selfe, nor by any other meanes except the tradition of Gods Church'. He concluded from this that the church was infallible. For 'if she be fallible, our beliefe of Scripture, and all verities contained therin, cannot be certaine, and infallible. We must therfore grant the true Church of Christ to be infallible, if we will maintaine Christian fayth to be certainly true'. It was the church which told us 'that the Scriptures we have are true Scriptures', said Hobbes's Catholic friend Sir Kenelm Digby in the same year, and he argued that we should assent to its decrees on other 'articles of faith' also, since 'they alike depend of the same authority'. Scripture, said Catholics, could not readily be understood by the average individual. To permit each private person to interpret the bible for himself – as Protestants did – was a recipe for spiritual and civil anarchy. Scripture, then, was to be interpreted not by the individual's 'private spirit' but by the church. Anyone who followed his own 'private sence' rather than 'the infallible iudgement of the church' would fall into 'pryde, the mother of heresy'.[5]

For many Protestants, reason and the testimony of the church might provide inducements to believe that certain books were

Scriptural. But the authority which confirmed such belief was the Holy Spirit working in the individual believer. Calvin gave rational grounds for thinking that the books in the bible were indeed God's word. But he insisted that what validated such belief was the operation of the Holy Spirit, working in the elect. Thomas Rogers argued similarly in his quasi-official commentary on *The Faith, Doctrine, and religion, professed, & protected in the Realme of England*, first published in 1608 and frequently reprinted. Rogers admitted that 'godly men in the Church' 'have, and doe receive' the books of the New Testament as canonical. But he claimed that 'we judge them Canonicall' less because others do so than because 'the holy Spirit in our hearts doth testifie that they are from God'.[6]

Some Protestants gave rather more emphasis to reason, and rather less to the Holy Spirit. Richard Hooker stressed the importance of rational argument in authenticating Scripture, and though he did not wholly exclude the role of the Spirit, he warned that it could be difficult to distinguish 'the Spirit of God' from 'the Spirit of error'. Chillingworth took the same line still further. He interpreted Hooker as meaning that 'natural reason,' 'built on principles common to all men' was the ultimate determinant of what constituted Scripture, and this was the position he himself adopted. In Chillingworth's opinion, the evidence for biblical events was of just the same kind as for any past actions. 'We have', he said, 'as great reason to believe that there was such a man as Henry the eighth King of England, as that Jesus Christ suffered under Pontius Pilate'. This rationalistic approach to Scripture was 'blasphemy', said Chillingworth's Calvinist critic Francis Cheynell, who insisted that when I read the bible God speaks 'to my heart and conscience', making Scriptural truths far more certain than anything which is based on mere fallible human testimony. Chillingworth was greatly influenced by Grotius, who defended Christianity on rational grounds in his *De veritate religionis Christianae* (1627). By de-mystifying Scripture, Grotius and Chillingworth attracted accusations of heresy – as did Hobbes.[7]

In answering the question of what constitutes Scripture, Hobbes stressed the importance of two connected distinctions – between law and counsel, and between obedience and belief. 'By the Books of Holy SCRIPTURE', he said 'are understood those, which ought to be the *Canon*, that is to say, the Rules of Christian life' (Lev 33: 260/199). However, 'There be two senses, wherein a Writing may be said to be *Canonicall*'. In the first sense, canons are rules 'given by a Teacher to his Disciple, or a Counsellor to his friend, without power to Compell him to observe them'. In the second sense, they are laws, 'given by one, whom he that receiveth them is bound to obey' (Lev 42: 356/281). Laws must be obeyed, but counsels need not be. If lawful authorities command us to profess that the Scriptures are the word of God, we are bound to do so; but this is not at all the same thing as saying that we are obliged to *believe* that the bible is indeed God's word. Human authorities are able to compel us to perform outward actions, but are not able to force us to believe anything. 'To obey is to do or forbear as one is commanded, and depends on the will; but to believe, depends not on the will, but on the providence and guidance of our hearts that are in the hands of God Almighty' (EW4:339; cf. Lev 42: 360/285). The ordinary means by which God makes men believe is through teaching: 'the means of making them beleeve which God is pleased to afford men ordinarily, is according to the way of Nature, that is to say, from their Teachers' (Lev 43: 406/323–4).

In order to rebut the thesis that we know the bible is the word of God because an infallible church has told us so, Hobbes employed the classic Protestant argument that the (Roman Catholic) church's claims to infallibility themselves rest upon Scripture. Against the notion that the Catholic church was the infallible guarantor of the Scriptural text, Chillingworth, for example, had argued that nothing can 'be pretended to give evidence to it, but only some places of Scripture; of whose incorruption more then any other what is it that can secure me? If you say the Churches vigilancy, you are in a Circle'.[8] 'How', asked Hobbes, 'shall a man know the Infallibility of the Church, but by knowing first the Infallibility of Scripture?'. So Catholic claims were viciously circular. But Hobbes equally rejected the contention of 'the other side' – of most Protestants – who based their belief that Scripture is God's word 'on the Testimony of the Private Spirit'. 'For how shall a man know his own Private spirit to be other than a beleef, grounded upon the Authority, and Arguments of his Teachers; or upon a Presumption of his own Gifts?' (Lev 43: 406/323). Hobbes joined with Catholics in rejecting the notion that God directly informs individual believers that Scripture is his word. At the same time, he joined with Protestants in rejecting the idea that the church is an infallible guardian of Scriptural truths.

Without direct revelation from God, he argued, no individual could *know* that the bible was indeed God's word. There remained the possibility, however, that there could be grounds for *believing* that this was the case. Those who lack direct revelation and yet believe in Christian doctrines, said Hobbes, take their beliefs from the teaching of people in whom they have faith. Consequently, belief that the bible is God's word rests upon the testimony of others; we believe this testimony because we have faith in those who give it to us. That is to say, we believe what they tell us because we also believe in their virtue and especially in their veracity (Lev 7: 48/31). It is important, Hobbes insisted, that we be wary of 'Ambition and Imposture' on the part of those who claim to tell us revealed, supernatural truths (Lev 36: 297/230), and he frequently warned against the ambition of churchmen. Yet there is good reason, he argued, to think that the clergy[9] have not in fact distorted the Scriptural message. Had they done so, we would expect Scripture to favour their own ambitious claims, but in fact it does not do this. 'I see not therefore', he concluded, 'any reason to doubt, but that the Old, and New Testament, as we have them now, are the true Registers of those things, which were done and said by the Prophets and Apostles' (Lev 33: 266/204).

So the bible was a reliable record of the words and doings of those humans who feature in its pages. Did Hobbes believe that it tells us anything about what God has said? On his own account, it is difficult to see what good grounds there can be for believing that God has supernaturally revealed anything to anyone. 'How God speaketh to a man immediately', asserted Hobbes, 'may be understood by those well enough, to whom he hath so spoken; but how the same should be understood by another, is hard, if not impossible to know'. Someone might indeed claim that he has been favoured with immediate divine revelation; but 'I cannot easily perceive what argument he can produce, to oblige me to beleeve it' (Lev 32: 256/196). It is therefore difficult to see what good grounds the earliest (or any subsequent) Christians could have for believing that the Christian message had been immediately revealed by God. True, Hobbes recorded that 'it is beleeved on all hands' that the 'first and originall *Author*' of the Scriptures 'is God' (Lev 33: 267/205). But it is difficult to show that he himself believed this – except, perhaps, in the irrelevant sense that God is the first author of everything. Hobbes shared much of the rationalistic approach to religion of Grotius and Chillingworth. But while they held that reason and historical evidence were sufficient to provide good grounds for belief in supernatural truths, it is unclear that Hobbes did so. '[R]eason', said Chillingworth, 'will convince any man, unless he be of a perverse mind, that the scripture is the word of God'.[10] For Hobbes, by contrast, reason will do no such thing. (These and related points are discussed in more detail in Chapter 6, section 2 below.)

God, said Hobbes, 'ought to be obeyed, whatsoever any earthly Potentate command to the contrary'. But subjects 'that have no supernaturall revelation' can have no access to God's commands except through reason, and reason tells them to obey their sovereigns. If the sovereign informs them that some book is the word of God, then they are not to deny this, though they may not believe it. 'According to this obligation', he said, 'I can acknowledge no other Books of the Old Testament, to be Holy Scripture, but those which have been commanded to be acknowledged for such, by the Authority of the Church of *England*' (Lev 33: 260/199). The books of the New Testament, he added, were 'acknowledged for Canon by all Christian Churches' (Lev 33: 261/200). He argued that subjects must refrain from publicly criticising the sovereign's interpretation of what the bible says. Arguably, he tried to abide by this precept in *De Cive*, which contains a number of passages that sound very like orthodox Anglicanism.

But when he wrote *Leviathan* he felt less need to tailor his professed opinions so that they complied with the sovereign's requirements. As he later said, he was not forbidden 'when I wrote my *Leviathan*, to publish anything which the Scriptures suggested. For when I wrote it, I may safely say there was no lawful church in England, that could have maintained me in, or prohibited me from writing anything'. 'There was then', he trenchantly commented, 'no church in England, that any man living was bound to obey' (*Answer to Bramhall* in EW4:355. Cf. also *An historical narration concerning heresy*, EW4:407). Indeed, in *Leviathan* itself he implied that there was no lawful authority in the country. For he declared that in dealing with controverted points of religion he intended only to *suggest* answers, and not dogmatically to maintain them, 'attending the end of that dispute of the sword, concerning the Authority, (not yet amongst my Countrey-men decided,) by which all sorts of doctrine are to bee approved, or rejected' (Lev 38: 311/241).[11] In other words, England was still at war, and until the war was over and a government established, it was open to private individuals to publish their own interpretations of Scripture. Moreover, the eccentricity of his doctrines on a great many points suggests that he was keen to flout rather than bow to orthodoxy. So there is little reason to doubt that the interpretation of Scripture offered in *Leviathan* represents Hobbes' own views on what the bible says, and was not doctored to accord with publicly instituted rules. That the same goes for *De Cive* is far less clear.

From Scripture, Hobbes drew conclusions about the relationship between church and state. There is no doubt that he thought the establishment of the truth on this question was of the highest importance in securing peace in England. In 1641 he told the Earl of Devonshire that the dispute 'betweene the spirituall and civill power, has of late more than any other thing in the world, bene the cause of civill warre'.[12] Events bore out his claim, for disagreements on questions of church government played a large part in bringing about civil war in England in 1642, and debate on this issue remained heated throughout the 1640s and 1650s. It is no accident that far and away the longest chapter in *Leviathan* is the forty-second – 'Of Power Ecclesiasticall'. In what follows we shall examine the relationship between Hobbes' ideas and those expressed by his contemporaries on church-state relations.

2: ECCLESIASTICAL POWER: THE CASE AGAINST BELLARMINE

Near the end of *Leviathan* Hobbes conveniently summarised his views on the history of church government in England. It is a story which starts with decline from pure beginnings. At first, Christians obeyed the Apostles because of their virtues, and 'out of Reverence, not by Obligation'. Then, as Christianity spread, the clergy began to meet together in assemblies to agree on what doctrines they should teach. They proceeded to make 'it to be thought the people were thereby obliged to follow their Doctrine, and when they refused, refused to keep them company'. This began the clergy's assault upon the liberty of the laity. It was, said Hobbes, 'the first knot upon their Liberty'. Later, as the number of presbyters (or priests) increased, those in the leading cities or provinces began to arrogate to themselves power over their fellows, and used the name of bishop to distinguish themselves from mere presbyters. This, said Hobbes, 'was a second knot on Christian Liberty'. Finally, the bishop of the most important city of all – Rome – acquired power over the other bishops, and the third and last knot was tied. This, in Hobbes's view, completed the process of decline. It was followed – many centuries later – by a gradual return to primitive purity, at least in England. This began with the untying of the third knot when 'the Power of the Popes was dissolved totally by Queen Elizabeth'. Nearly eighty years afterwards the second knot was untied when 'the Presbyterians lately in England obtained the putting down of Episcopacy'. Then the first knot was also dissolved, for power was taken from the Presbyterians. 'And so', concluded Hobbes, 'we are reduced to the Independency of the Primitive Christians', where all were free to follow whatever minister they pleased. This system of church government, he remarked, 'is perhaps the best' (Lev 'Review and Conclusion' 479–80/384–5).

Hobbes held that the ideas of Presbyterians and Anglicans included objectionable elements. But he regarded Roman Catholic notions on ecclesiastical power as the worst of all. The bulk of what he had to say against false theories of clerical power was primarily directed against Catholics, and especially against Cardinal Robert Bellarmine – whose works are quoted far more extensively in *Leviathan* than any other writings except the bible. Hobbes treats Presbyterian and Anglican views on the clergy's power as modified versions of Bellarmine's theory, arguing that they incorporated some but not all of its errors. So a refutation of Bellarmine would simultaneously demolish the mistaken opinions of Presbyterians and Anglicans. Hobbes later claimed that in *Leviathan* he 'suffered the clergy of the Church of England to escape' from criticism (*Six lessons* in EW7:354). But though his arguments were ostensibly aimed at Bellarmine, many of his incidental comments were clearly targetted as much against Anglican or Presbyterian foes.

By attacking Bellarmine, Hobbes was able to place himself in a venerable tradition of Protestant polemic – for Bellarmine's views had already been criticised by scores of Protestant writers. As we have seen, the Cardinal played a major part in two great controversies over church-state relations which broke out during Hobbes' Oxford days. One related to the Interdict placed on Venice by Pope Paul V, while the other was concerned with the Jacobean oath of allegiance. It is highly likely that Hobbes was familiar with what was said in both of these debates. Certainly, his arguments often coincide with those employed by the Venetians or by James I's supporters. 'The papal monarchy', said Pierre Du Moulin in a reply to Bellarmine which was published in 1614 and which was in the library of Hobbes' patron,[13] 'was born from the ruins of the Roman Empire'.[14] Hobbes, as often, put the thing better but did not really say anything very new when he declared in *Leviathan* that 'the *Papacy*, is no other, than the *Ghost* of the deceased *Romane Empire*, sitting crowned upon the grave thereof' (Lev 47: 480/386).

Like the defenders of James I's oath of allegiance, Hobbes argued that the pope possesses no temporal power outside the papal states, and used Scripture to support his case. The High Priest Jehoiada, he said, deposed Queen Athaliah not in his own right but in that of the true ruler, Joash. So the bible gives no power over kings to priests – or popes – acting on their own authority (Lev 44: 426/341; cf. 42: 402/320). Jacobean writers argued in precisely the same way. King Solomon, declared Hobbes, had deposed the High Priest Abiathar, and he concluded that monarchs have the authority to discipline clerics (Lev 40: 329/254; 42: 394/313; 44: 426/341). Others said the same thing. St Ambrose's action in excommunicating the Emperor Theodosius – 'if it were true hee did so' – was criminal, said Hobbes (Lev 42: 402/320). James I's supporters likewise claimed that sovereigns cannot licitly be excommunicated and that St Ambrose had either acted wrongly or else had not really excommunicated the emperor. The term pastors applies not only to clerics but also to sovereigns, said Hobbes (Lev 42: 372-3/295-6). Andrewes and Burhill had made the same point – as had Grotius.[15]

The essence of Bellarmine's case in favour of the thesis that the pope has the authority to depose sovereigns rested upon two connected propositions. The first was that the Roman Catholic church is a perfect community or commonwealth, possessing full political rights including the right to make war on other commonwealths. As the leading English Catholic Matthew Kellison put it, the church is an 'absolute commonwealth' empowered to 'use defensive war, not only against Christian princes, but also even against pagans and infidels that do molest her'. Where appropriate, he argued, the church could also wage an aggressive war.[16]

The second proposition was that the church exists to promote the spiritual welfare of Christians, and that since spiritual goals are to be preferred to merely temporal ends, the church (acting through its representative the pope), may intervene in the temporal affairs of states whenever His Holiness deems that such action would promote the spiritual good. Because spiritual goals or ends are superior to temporal ones, so the argument ran, the spiritual power of the ecclesiastical commonwealth is superior to the temporal power of the state. 'Temporals should be ordered to a spiritual end', said Suarez, and he concluded that 'just as the ends are subordinated, so are the powers'. This theory was commonplace though by no means universal amongst Catholics. Its proponents claimed that the pope was not the direct temporal overlord of Christian kings, but that in exercising his spiritual powers he could indirectly intervene in the temporal affairs of kingdoms – for example by deposing monarchs. His power in temporal matters, they said, was not direct but indirect.[17]

During the controversies over the Venetian Interdict and the English oath, some standard arguments were employed by critics of Bellarmine to rebut the idea that the pope may ever exercise temporal dominion outside his own central Italian territories. These arguments also feature in Hobbes' writings. The distinction between direct and indirect papal power in temporal matters was attacked as a mere play on words. Whether the pope claims to be acting for temporal or spiritual ends in deposing a monarch 'the effect is the same' and the power itself is temporal, said William Barclay; while Du Moulin argued that to tell a king that he was being deposed by an indirect rather than a direct temporal power was like comforting a man on the scaffold by letting him know he would be beheaded with an ax rather than a sword. 'Power is as really divided and as dangerously to all purposes', said Hobbes, 'by sharing with another an *Indirect* Power, as a *Direct* one' (Lev 42: 396/315). The argument from the subordination of temporal to spiritual ends was criticised as inadequate. The fact that we ought to prefer spiritual to temporal goals, it was said, by no means implied that spiritual was superior to temporal power. The goals of cookery, Du Moulin observed, are subordinate to those of medicine, for prudent cooks will take account of the precepts of doctors in exercising their art; yet no one thought that doctors should hold power over cooks. Hobbes admitted that the sovereign 'ought indeed to direct his Civill commands to the Salvation of Souls', but he denied that kings were subject to the pope (Lev 42: 398/316; EL2:9:2). The 'art of a Sadler', he said, is subordinate 'to the art of a Rider' but it by no means follows that 'every Sadler is bound to obey every Rider'. Similarly, even if the pope does possess spiritual power, it cannot be concluded that princes are subject to him (Lev 42: 396–7/315).[18]

Opponents of Bellarmine's theory denied that the church is a kingdom or commonwealth independent of civil states. Bellarmine and like-minded Catholics were not, of course, claiming that the church is an ordinary temporal monarchy, set up by people according to the principles of natural law. Rather, their argument was that Christ, acting outside the ordinary course of nature, had possessed spiritual sovereignty on earth and that he had passed this sovereignty on to St Peter, who in turn had transmitted it to his successors, the popes. So the church was a spiritual monarchy, governed by the pope.

One vulnerable aspect of this argument was that its Scriptural foundations were open to question. A crucial text was Matthew 16:18–19, where Christ says '*Thou art Peter, And upon this rock I will build my Church*' (Lev 42: 379/301). Bellarmine's critics argued that in these words Christ grounded the church not upon the person or papal office of St Peter, but upon his faith, and Hobbes took the same line (Lev 42: 380/301). They also claimed that Scripture gave no support to the notion that the church was a monarchy independent of civil states and wielding power over all Christians. Christ had not laid claim to kingly power while he was on earth, said Bishop Buckeridge, but had specifically stated that 'My kingdom is not of this world'. So Christ could not have passed monarchical authority on to Peter or his successors, and papal pretensions were therefore unwarranted. Christ, asserted Hobbes, 'expressly saith, (*John* 18. 36.) *My Kingdome is not of this world*' (Lev 41: 333/262; cf. EL2:7:9). He drew the straightforward conclusion that Christ's ministers could not exercise kingly power in his name – unless they happened to be kings anyway (Lev 42: 341/269).[19]

Anglicans commonly held that in any Christian country, church and commonwealth are simply terms describing two different aspects of a single community. The church of England was the English people viewed as Christians, while the commonwealth was the same people regarded as a temporal community. In a Christian country, said Richard Hooker, '[t]he *Church* and the *Commonwealth* . . . are . . . one societie . . . being termed a *Commonwealth* as it liveth under whatsoever forme of secular lawe and regiment, a *Church* as it hath the *Spirituall* lawe of *Jesus Christ*'. He went on to argue that 'where one and the self same people are the *Church* and the *Commonwealth*', natural reason indicated that 'their Soveraigne *Lord* and Governour in causes civill' should 'have also in *Ecclesiasticall* affayres a supreme power'. A Christian monarch, said Anglicans, was the supreme governor or head of the church in his realm; Hobbes adopted exactly the same position. 'That which is called a city in so far as it is made up of men', he declared, 'is termed a church in so far as it consists of Christians' (DC17:21: 'Quae vero civitas vocatur, quatenus conflatur ex hominibus, eadem quatenus constat ex Christianis, Ecclesia nominatur'). A church, he said, 'is the same thing with a Civil Common-wealth, consisting of Christian men'. The supreme governor of the church, he concluded, is the sovereign (Lev 39: 321–2/248).[20]

There were some novel elements in Hobbes' use of Scripture against Bellarmine. For example, he argued that in the period between the death of Moses and the appointment of Saul as king, Israel had been ruled by High Priests. Usually it was papalists who claimed that the High Priest had ruled, for they viewed him as a forerunner of the pope. The Jesuit Thomas Fitzherbert, for instance, asserted that the High Priest had had 'a sovereignty of authority'. Most of Bellarmine's critics, by contrast, argued that Moses was succeeded by Joshua but that thereafter, though there were officers and judges in Israel, there was no chief governor until the time of Saul. The High Priests, they said, had certainly not held sovereign power, and Eleazar had been Joshua's subject.[21] Hobbes, however, insisted that Joshua had been subject to Eleazar, who had held sovereign authority, as did his successors in the office of High Priest up until Saul's time (Lev 40: 327/253). By reading Scripture in this way, Hobbes was able to endorse Bellarmine's lofty claims for the powers of the High Priest but draw conclusions very different from the Cardinal's. The High Priest held his powers, Hobbes claimed, in virtue of his office as a civil sovereign. So Bellarmine's examples showed that there was biblical testimony that sovereigns possess these powers, but proved nothing at all about the pope (Lev 42: 385/305–6, 386/307).

On a wide variety of points, however, Hobbes took just the same line against Bellarmine as had earlier thinkers; he rehearsed the standard examples of Athaliah, Solomon and Abiathar and so on, and argued that kings as well as clerics are pastors. He claimed that the subordination of temporal to spiritual ends implies no subordination of monarchs to popes or priests. He held that the pope possesses no monarchical authority over Christians, pointing out that Christ had said His kingdom is not of this world and that He had in any case founded the church not upon St Peter's person but upon his faith. All of these arguments were perfectly conventional, and can easily be paralleled in scores of treatises against Catholics. Yet *Leviathan* and to a lesser extent *De Cive* attracted bitter criticism from Anglicans.[22] For Hobbes did not confine himself to attacking Bellarmine, but also set out his own positive conclusions on church government, and in the process attacked some of the central contentions of orthodox Anglican thinking.

3: ECCLESIASTICAL POWER: *DE CIVE, LEVIATHAN* AND ANGLICAN THINKING

Protestants denied that clerics possess any power in temporal matters except that which the civil ruler chooses to grant them. Puritans frequently added that he could not or should not give them such power, since their spiritual duties are sufficiently onerous to occupy all their energies and they might abuse temporal power to further their ambitions. Hobbes held that a cleric has no temporal power – unless he happens to be a sovereign, or to hold a commission from one. He did not adopt the puritan view that such commissions are sinful, and he even asserted that the sovereign may delegate authority to the pope himself (Lev 42: 378/300). Anglicans granted the sovereign the power to further not only the temporal but also the spiritual good of his subjects. The monarch, they said, is the supreme governor of the church, and has a duty to make laws and to enforce them by coercive measures in order to promote spiritual welfare. Again, Hobbes had no quarrel with this position.

But Anglicans went on to argue that the ruler should govern the church through the clergy. Clerics, they claimed, possess certain spiritual powers which they derive from God alone and not from the sovereign. For example, they said that the clergy are empowered to preach (and in preaching to interpret Scripture), to administer the sacraments, to excommunicate sinners, and to create new clerics by the imposition of hands in ordination or consecration. The sovereign, they asserted, may do none of these things, but he alone can make laws and employ coercion to ensure that the clergy perform their functions in the best interests of the community. So the clergy were to govern the church by employing their spiritual powers, while the monarch used coercive power to ensure that clerics carried out their duties properly and to back up the clergy's actions. Clerics, they said, derived their spiritual powers from God alone through ordination or consecration, but their exercise of these powers in any Christian country was subject to the sovereign's laws.[23] Many argued that by divine right bishops had greater powers than other clergy.[24] Many also claimed that God Himself has decreed that the clergy be funded by tithes – ten per cent of the income of the laity.

Hobbes had much to say on all these topics. In important respects, the case he mounted in *De Cive* seems to differ markedly from that which he put forward in *Leviathan*.

In *De Cive*, Hobbes argued that the sovereign alone is to decide all questions which belong to the sphere of natural reason. 'But to decide on questions of faith – that is to say, concerning God – which transcend human abilities, there is need for a divine blessing (so that we may not err, at least on necessary points) to be derived from Christ Himself by the imposition of hands' (DC17:28: 'Quaestionibus autem fidei, id est, de Deo, quae captum humanum superant, decidendis, opus est benedictione divina (ne possimus falli, saltem in necessariis) per impositionem manuum ab ipso Christo derivanda'). Reason alone, he argued, could not show us the way to salvation; we therefore require supernatural information on this matter. Christ, he proceeded, has promised infallibilty of judgement on questions of faith to the Apostles and to those who have succeeded them by imposition of hands – that is to say, the clergy. So Christian sovereigns are obliged to employ 'properly ordained ecclesiastics' ('Ecclesiasticos rite ordinatos'; DC17:28) in deciding on questions of faith. There is nothing in *De Cive* to suggest that sovereigns themselves can preach or baptise. Moreover, in the second and later editions of the book Hobbes clearly distinguished between bishops and other clerics, asserting that not all presbyters (or priests) had been bishops in the time of the early church (DC17:24; cf. EL2:7:8). All this looks very much like standard Anglican doctrine. It is strikingly different, at least in appearance, from what Hobbes says in *Leviathan*.

The essence of the usual Anglican position was that clerics possess some purely spiritual powers which they derive directly from God, but that these powers in no way conflict with the monarch's temporal sovereignty or with his supremacy over the church. For the clergy's powers were not temporal, nor could they be exercised except according to the sovereign's laws. The essence of Hobbes' teaching in *Leviathan*, by contrast, was that clerics have no powers whatever, except those which the sovereign chooses to grant them. Christ's kingdom, he observed, was not of this world. So His ministers held no authority on earth. Their office was not to command but to persuade (Lev 42: 341/269, 361/286). Spiritual *government* was simply a misnomer, for '[t]here is . . . no other Government in this life, neither of State, nor Religion, but temporal' (Lev 40: 322/248). The notion that there is spiritual as well as temporal government was a concoction of ambitious clerics, intended only to further their own interests: '*Temporall* and *Spirituall* Government, are but two words brought into the world, to make men see double, and mistake their *Lawfull Soveraign*' (Lev 40: 322/248).

In *Leviathan*, Hobbes pursued the implications of these ideas relentlessly, demolishing the Anglican theory along the way. 'All Pastors, except the Supreme', he declared, 'execute their charges in the Right, that is by the Authority of the Civill Soveraign'. Bishops derive none of their powers directly from God, and in claiming to do so they 'sliely slip off the Collar of their Civill Subjection, contrary to the unity and defence of the Common-wealth' (Lev 42: 374/296–7). Ecclesiastics, he argued, have no independent jurisdiction at all, for 'Jurisdiction is the Power of hearing and determining Causes between man and man; and can belong to none, but him that hath the Power to prescribe the Rules of Right and Wrong; that is, to make Laws; and with the Sword of Justice to compell men to obey his Decisions' (Lev 42: 391–2/311). This was, of course, the sovereign. Since no cleric holds authority except by commission from the sovereign, it follows that before the conversion of rulers to Christianity, clerics were wholly lacking in authority. But if none held any authority, there could manifestly have been 'no diversity of Authority' amongst them (Lev 42: 365/289). The Anglican contention that bishops had wielded authority over other pastors in the early church was therefore false, and Hobbes declared that '*Bishop, Pastor, Elder, Doctor,* that is to say, *Teacher,* were but so many divers names of the same Office in the time of the Apostles' (Lev 42: 365/289).

In *De Cive* Hobbes asserted that Christian sovereigns are obliged to employ properly ordained clerics in interpreting Scripture. In *Leviathan*, this provision disappeared. Moreover, Hobbes now insisted that the sovereign may perform all the other functions traditionally confined to clerics, including preaching, administering the sacraments, and consecrating 'both Temples, and Pastors to Gods service' (Lev 42: 374/297). The only reason why the sovereign did not normally do these things was that 'the care of the summe of the businesse of the Common-wealth taketh up his whole time' (Lev 42: 375/297). By adopting this position Hobbes diverged not only from Catholics, Anglicans and Presbyterians, but also from Independents. For all of these groups held that sovereigns are not empowered to perform the functions of clerics.

Anglicans argued that it was ordination by existing clergy which qualified a man to act as a minister. Of course, sovereigns had not normally received ordination; but Hobbes denied that this mattered. *De Cive* distinguished between the election of a pastor, which took place on the authority of the church (and therefore of the sovereign in a Christian commonwealth), and his ordination or consecration, which was to be carried out by clerics through the imposition of hands (DC17:24). It was by means of ordination, he said, that pastors received their infallibility in matters of faith (DC17:28). In *Leviathan*, however, he argued that ordination by imposition of hands was simply a ceremony by which the person elected as a pastor was 'designed' or pointed out. '[T]o design a man, or any other thing, by the Hand to the Eye, is lesse subject to mistake, than when it is done to the Eare by a Name' (Lev 42: 376/298). In other words, the purpose of laying on hands was to prevent any confusion about who had been elected. Since the sovereign's person was obvious, it did not need to be 'designed'. The Apostles, indeed, had given newly elected pastors 'the benediction, which is now called Consecration' (Lev 42: 366/290); but this did not make them pastors, nor grant them any powers.[25]

The idea that the sovereign must interpret Scripture by means of properly ordained pastors is wholly absent from *Leviathan*. A few years after its publication, Hobbes returned to the same theme, inveighing against Wallis's contention that ministers must be ordained by other pastors. '[T]he right of ordination of ministers', he declared, 'depends not now on the imposition of hands of a minister or presbytery, but on the authority of the Christian sovereign'. If the sovereign chose, he could appoint any person – or at least any man[26] – to preach or perform other ecclesiastical functions. '[M]inisters of the Gospel', he said, 'are those to whom the preaching of the Gospel is enjoined by the sovereign power in the name of Christ'. He added that he himself would be a minister if the sovereign commanded him 'though without the ceremony of imposition of hands, to teach the doctrine of my *Leviathan* in the pulpit' (EW7:397). It is unclear that Hobbes wanted to be a minister, though he certainly thought that the doctrines of *Leviathan* should be publicly taught.

In *De Cive* Hobbes stayed silent on the question of how clerics were to be financed. In *Leviathan*, however, he did not – and what he had to say was once again ill-calculated to please orthodox Anglicans. During the later sixteenth century the notion that by God's law clerics should receive tithes had rarely featured in the writings of conformist English divines. But from the first decade of the seventeenth century they commonly argued that tithes were indeed due to the clergy by divine right.[27]

The classic reply to this claim was John Selden's *Historie of Tithes*, published in 1618. Selden did not quite deny that tithes are prescribed by God's law; he just showed that where tithes had in fact been paid, the civil authority had decreed that they ought to be. The obvious implication was that clerical finance is subject to the sovereign. Selden cautiously refrained from spelling this out, though James I – acting on the advice of clerics – nonetheless forbade him from replying to attacks on his book. In *Leviathan* Hobbes adopted Selden's line, and displayed no qualms at all about thrusting home its implications – for he roundly attacked the idea that tithes are due to the clergy by divine right (Lev 44: 421/336).

In ancient Israel the tribe of Levi received a tenth of the fruits of the land from the other tribes, said Selden and Hobbes, and the priests – who were Levites – a tenth of that tenth (Lev 42: 369/292). This arrangement was instituted 'by the Civill Power', declared Hobbes (Lev 42: 369/293), and the same conclusion is readily deducible from Selden's account. In the time of the Apostles, asserted Hobbes, 'the Christians of every City lived in Common' – and therefore not on tithes. He cited Acts 4:34 to prove the point (Lev 42: 370/293). '[T]he whole Church, both Lay and Clergie, then livd in common', said Selden, citing Acts 4:34. In those days, declared Hobbes, tithes were not paid to Christian pastors, who lived on the voluntary contributions of the faithful. This did not mean that clerics lived in poverty, he added, and he cited the fourth-century historian Ammianus Marcellinus to substantiate his thesis that Christians – and 'especially . . . Matrons' – were often very generous towards their pastors (Lev 42: 371/294). Until about the year 400 clerics had lived on the voluntary contributions of their flocks, said Selden. He cited the same passage from Ammianus Marcellinus upon which Hobbes was later to draw. Clerics, concluded Hobbes, are to live either upon the benevolence of their flocks or upon whatever maintenance the sovereign chooses to grant them (Lev 42: 371/294).[28]

De Cive seemingly differs considerably from *Leviathan* on questions of church-state relations. In the earlier book, Hobbes suggests that pastors must be ordained by imposition of hands in a chain of ordinations traceable back to the Apostles. He says nothing about tithes and nothing to imply that sovereigns may exercise the functions of the clergy. All this looks very different from the argument of *Leviathan*. What exactly is the nature of these differences, and why did Hobbes change his mind, or at least his tune?

According to Robert Payne, Hobbes had special reasons for being irritated with the Anglican clergy when he wrote *Leviathan*, for he believed that some of them had undermined his position in the good graces of Charles II. This may well help to explain the passion of some of the anti-clerical passages in the book; but the basic theory on church-state relations expressed in *Leviathan* is the same as that put forward in *De Cive*. Arguably, Hobbes' teaching on bishops and ordination in the earlier work is inconsistent with that theory, and was tacked on in order to conform to the sovereign's wishes or to avert royalist criticism.

As we have seen, in the summer of 1641 Hobbes wrote to the Earl of Devonshire, commenting that recent bishops had been guilty of many abuses in the way they had run the church, and approving of proposals to replace them with lay commissioners. All powers of church government, he added, were derived from the secular authorities, so the sovereign might abolish episcopacy and use laymen to manage ecclesiastical affairs.[29] A few months after writing this letter, Hobbes completed the manuscript of *De Cive*. In this, and again in the first printed edition of the book, he drew no distinction at all between bishops and other clerics. In what was clearly a royalist work, this was highly surprising. Charles I's own devotion to episcopacy as God's ordinance was well-known, and it proved to be a principal reason for his failure to reach terms with his opponents. In the second (and first widely circulated) edition of *De Cive* Hobbes gestured in the direction of conventional royalism by asserting that not all presbyters had been bishops in early Christian times (DC17:24), but he mounted no arguments to support this contention, and its significance is far from clear. Standard royalist doctrine maintained that by God's law bishops differed in office and authority from other pastors, but *De Cive* says nothing about the nature of the difference. That Hobbes then thought bishops held greater powers than other pastors is doubtful, for the implication of much of his argument is that no cleric holds any powers.

Virtually all thinkers claimed that the doctrines which Christ had revealed to the faithful are (at least in part) laws. Christ might not, indeed, have ruled as a king on earth, and perhaps did not Himself legislate. But He did reveal His Father's laws for Christians. Of these laws, some defined the nature and powers of the church –

for instance instituting the clergy as the appropriate interpreters of Scripture, and spelling out what functions belonged to a bishop's office. In *De Cive*, however, Hobbes held that Christ declared only old laws, which had already been promulgated. All the laws which Christ endorsed, he argued, were either natural or else had been introduced by men who held sovereignty over the people of Israel – and who also, as it happened, told their subjects that the laws in question had been directly communicated to them by God Himself (DC17:8; cf. 16:3–18). Since none of these laws deal with the powers of Christian clerics, it follows that there are no laws at all on the rights of the clergy – unless the sovereign happens to introduce them. So it makes no sense to claim that the sovereign 'is obliged' ('obligatur') to interpret Scripture by means of properly ordained clerics (DC17:28). Yet Hobbes made just this claim.

There are other problems with what Hobbes says in *De Cive*. He asserts that Christ promised infallibility on matters of faith to the Apostles and their successors (DC17:28), yet elsewhere argues that whoever regards a man as infallible will obey him in all things. To claim infallibility, then, is to claim dominion (DC18:14) – and Hobbes had no wish to grant dominion to churchmen. Perhaps he had this point in mind when he asserted that the Apostles were infallible only on matters of faith (DC17:28), and that it is the civil sovereign alone who decides what constitutes such matters – for 'it belongs to temporal right to define what is spiritual, and what temporal' (DC17:14: 'quid sit spirituale, quid temporale . . . pertinetque ad ius temporale'). This principle effectively destroys the clergy's infallibility, for the sovereign can always declare that their pronouncements pertain to temporal matters, and not to faith. Moreover, on Hobbes' own account, if the sovereign ignores the clergy and interprets Scripture either by himself or with the aid of lay commissioners, his subjects must publicly adhere to his interpretation – provided only that he does not command them to deny or insult Christ (DC18:13; cf. also Hobbes' later discussion of infallibility in EW5:269).

Already in *De Cive*, then, Hobbes argues that if the sovereign is a Christian, we are to obey him in all matters, both spiritual and temporal. But where the ruler is a heathen, he tells us, we are to follow 'some Christian church' ('Ecclesia aliqua Christianorum') on spiritual questions; if the heathen prince is displeased at this, we are not to resist him but cheerfully to accept martyrdom (DC18:13). In *Leviathan*, Hobbes modified this doctrine in two respects, arguing that before the conversion of sovereigns to Christianity all individuals were at liberty to interpret Scripture for themselves (Lev 42: 355–6/281), and also that they could publicly deny Christ if commanded to do so by the ruler (Lev 42: 343–4/271, 414/331; Hobbes excepts the clergy from this provision: Lev 45: 452/362; *An answer to Bishop Bramhall's book*, EW4:321).

The doctrine of *De Cive* that the faithful in a pagan state must follow 'some Christian church' is problematical for several reasons. Firstly, it is hard to see what law there can be which obliges anyone to follow a body other than the commonwealth. For natural law requires no such thing, and Hobbes argued that Christ introduced no new law. Secondly, Hobbes defines a church as a body which may be lawfully called together by someone, and whose members are bound to attend when summoned (DC17:20). A church, in short, is simply a commonwealth of Christians – for outside a commonwealth people are a disunited multitude which cannot be lawfully called together (DC17:20–1; cf. Lev 39: 321/247–8). Elsewhere in *De Cive* Hobbes does refer to various Christian churches which seemingly acted as united bodies in pagan times – such as those at Antioch (DC17:24), Rome and Jerusalem (DC17:26) – and in *Leviathan* he equates 'the Church' in 'the . . . times before the conversion of Kings' with 'the Assembly of Christians dwelling in the same City; as in Corinth, in the Assembly of the Christians of Corinth' (Lev 42: 349/275–6). But the principles Hobbes spells out elsewhere make it difficult to see what authority these assemblies could have had over anyone. In the third chapter of his Appendix to the Latin *Leviathan* he thrust home the point that under pagan emperors Christian assemblies had no authority whatever (LW3:566).

Moreover, Hobbes' theory implies that when sovereigns did not simply ignore but actually prohibited Christianity, such assemblies were unlawful and seditious conventicles.[30] In *Leviathan* he argued that there could be no authorised interpreters of Scripture 'till Kings were Pastors, or Pastors Kings' (Lev 42: 356/281), and deduced the consequence that before that time everyone was free to interpret the bible for himself. In *De Cive* he was keen to emphasise the anarchic consequences of granting individuals freedom to interpret the bible for themselves (DC17:27) and so required them to follow 'the church'. From the perspective of Hobbes' system, however, the 'church' before the conversion of sovereigns was an alarmingly nebulous entity. It still features in *Leviathan* – where, as we shall see in the next section, Hobbes exploits it to his own purposes – but individuals are no longer required to follow it.

Payne was right to think that Hobbes made his hostility to notions of divine right episcopacy explicit in *Leviathan*, though not in *De Cive*. On this point, *De Cive* was certainly the more conventionally royalist of the two works; but its Anglicanism was skin-deep, and on most of the fundamental questions relating to church government its teachings were close to those of *Leviathan*. The precept that sovereigns are obliged to interpret Scripture by properly ordained clerics does not cohere with the rest of what Hobbes says in *De Cive*: there is no law requiring sovereigns to do this, and he does not even provide us with a piece of biblical advice counselling them to such action. Advice like that would in any case not be part of the fundamental Christian message, and so could be ignored. The notion that under a pagan sovereign we must follow some Christian church in spiritual matters is equally problematical, since the nature and authority of such a church are obscure, and the pagan ruler must in any case be our guide on which matters are temporal and which spiritual. *De Cive* requires that subjects obey the sovereign – whether pagan or Christian – in all matters that he declares to be temporal. This doctrine, too, is very different from the usual teaching of English royalist churchmen – who believed that there were objective Christian laws which all must obey irrespective of what the sovereign said. Also very different from standard Anglican attitudes was the account which Hobbes gave of the most important power traditionally exercised by clerics – the power of excommunication. His position on this question invites comparison with that of Thomas Erastus – and his broad doctrine of church-state relations needs to be examined against the background of the Erastianism of his contemporaries.

4: EXCOMMUNICATION AND ERASTIANISM

Anglicans and other Protestants denied that churchmen might lawfully use temporal coercion except upon commission from the sovereign. But they generally granted ecclesiastics certain spiritual powers which they could use to encourage their flocks to refrain from sin and to do the Lord's work. The crucial power was excommunication. In England, ecclesiastical law recognised two forms of excommunication. The lesser suspended offenders from 'the use of the sacraments and divine worship'. The greater deprived them not only of these things, but also 'of the society and conversation of the faithful'.[31]

In England before the Reformation, the first type of excommunication was more common, but by the 1560s it had nearly disappeared. The church courts continued to use greater excommunication, and an act of Parliament of 1563 facilitated the arrest and imprisonment by the secular authorities of contumacious excommunicates. But so many people were excommunicated, and the cost of having them arrested would have been so great, that a good many went largely undisturbed.[32] With the fall of the church courts after 1640, ecclesiastical discipline all but vanished. Presbyterians hoped to replace episcopacy with their own form of church government, and many – particularly amongst churchmen – were keen to institute effective disciplinary measures to improve the morals and the religious knowledge of the population. Calvin had suggested that the principal means by which discipline should be enforced was through the exclusion from the Sacrament of the Lord's Supper of those whom the consistory of elders pronounced unworthy.[33] Many English Presbyterians lobbied for the introduction of these arrangements during the 1640s.

In the 1560s efforts to set up a similar system in Heidelberg had aroused bitter debate. A physician named Thomas Erastus produced a set of theses attacking the whole idea that churchmen acting independently of the civil magistrate can exercise powers of excommunication. He soon became involved in controversy over this question with Calvin's successor at Geneva, Theodore Beza. Eventually, Erastus himself was excommunicated. Some of his writings were published at London in 1589 as part of the campaign which the bishops mounted against Presbyterianism in the years after the Spanish Armada (and Hobbes' birth). During the 1640s, the works of Erastus provided much useful ammunition for critics of the Presbyterians – including Thomas Coleman, William Prynne, and John Selden. Hobbes' general argument was close to what such authors said, but on many details he differed from them, and from Erastus.[34]

It is sometimes said that the Anglican clergy was Erastian in outlook. This is true in so far as clerics granted the monarch supremacy over ecclesiastical affairs, and held that churchmen can exercise their functions only with their sovereign's permission. But on other points they parted company with Erastus. Shortly after the London publication of Erastus' works, the important Anglican theorist Hadrian Saravia wrote a letter to Richard Bancroft, later Archbishop of Canterbury. Saravia sharply criticised Erastus' doctrine that ecclesiastical censures should be abolished. Elsewhere he vigorously attacked the 'strange opinion' that 'the whole censure of manners is to be set over to the magistrate'. In Saravia's view, the clergy possesses the right to judge and discipline sinners. They might be accountable to the prince for how they exercise this power, but the power itself is granted to them directly by God, and does not depend upon the ruler. This was the conventional Anglican position. The power of excommunicating, said Sanderson, was 'part of the power of the keys which our Lord Jesus Christ thought fit to leave in the hands of his Apostles and their successors'. Kings held no such power, though they could regulate how clerics used it. 'The power which Christ hath given to the officers of his church', said Stillingfleet, 'doth extend to the exclusion of contumacious offenders from the privileges which this society enjoys'. Clerics excluded sinners from the church, though they were accountable to the king for how they exercised this power.[35]

The jurisdiction which churchmen exercised in excommunication was commonly distinguished from coercive civil jurisdiction – or jurisdiction strictly so called. The sentence of excommunication notified a sinner that if he failed to mend his ways he would not attain salvation, but excommunication did not itself deprive anyone of salvation – for it was God not man who saved. A main purpose of excommunication, it was said, was to encourage believers who had fallen into sin to reform their lives. But if the person excommunicated believed that he had not in fact sinned, or that those who pronounced the sentence of excommunication were incompetent to do so, he would be unlikely to change his lifestyle. So excommunication achieves its primary purpose only in the case of those who are willing to be persuaded that the sentence pronounced against them is just. Temporal jurisdiction, by contrast, operates upon us regardless of our thoughts about the magistrate. Again, as Taylor pointed out, if a sentence of excommunication is in fact unjust, it will not deprive the excommunicate of salvation; but 'in proper jurisdiction it is otherwise: for, right or wrong, if a man be condemned he shall die for it; and if he be hanged he is hanged'.[36]

A second purpose of excommunication – besides encouraging moral reform in sinners – was to tell the faithful that there was a wicked man amongst them so that they could avoid contact with him. It was held that believers should refuse to eat with anyone who had been excommunicated. An excommunicate, said Thomas Rogers in his exposition of the Thirty-Nine Articles, is 'to be avoided: and not to be eaten with all, not to be companied with all, nor to be receaved into house'. Catholics argued that there were certain circumstances in which this injunction did not apply, and in which people might sinlessly associate with excommunicates. Wives, servants and sons could continue to consort with excommunicate husbands, masters or fathers, and some thought it probable that you could sleep with an excommunicate if you were assigned the same bed in an inn.[37] Anglicans often extended the list of such cases to include subjects and sovereigns, arguing that the subject was obliged to continue to obey an excommunicated prince, and indeed that the sovereign could not properly be excommunicated at all.[38] While Bellarmine claimed that spiritual jurisdiction was vested in the pope as the successor of St Peter and that bishops could exercise it only as the pope's delegates, his Anglican critics argued that it was vested in the whole church, by which they meant the Apostles and their successors the bishops.[39]

Anglicans occasionally cited Erastus to support their arguments against Presbyterianism,[40] but they dissented from many of his central contentions. For Erastus asserted that excommunication is a modern invention, introduced to further the ambitions of clerics, and lacking any scriptural authority. The bible, he said, mentions only corporal penalties and only secular jurisdiction. He claimed that Christ's famous words in Matthew 18:15–17 provided no basis for excommunication. In that passage, Christ said that if your brother commits an offence against you, you should first remonstrate with him privately. If that does not work, you should try the same thing again in the presence of witnesses. If that also fails, you should 'tell it unto the church', and 'if he neglect to hear the church, let him be unto thee as an heathen man and a publican'. What this meant, said Erastus, was that you should set the case before the Jewish civil magistracy or sanhedrin – which is what the word church here signifies. If the offender refuses to accept the sanhedrin's jurisdiction, then you could take the case before a Roman tribunal, as you would if he were not a Jew but a gentile.[41]

Scripture, claimed Erastus, did not ban heathens from attending Christian church services.[42] It did indeed record that St Paul delivered sinners to Satan (1 Corinthians 5:5), but this was an extraordinary or miraculous power which was not passed on to later churchmen.[43] To allow two separate jurisdictions over conduct to exist in one commonwealth, he insisted, would spell anarchy.[44] If the magistrate was a Christian, his civil jurisdiction was quite sufficient to correct evil manners. Under a heathen ruler, Christians might voluntarily agree to set up arbitrators in order to settle disputes amongst fellow-believers. But such arbitrators would have no power to debar people from the sacrament of the Lord's Supper – or, in other words, to inflict the sentence of lesser excommunication upon them.[45]

We are now in a position to turn to Hobbes' teachings on excommunication and to assess their originality. Hobbes showed no interest at all in lesser excommunication, which he does not mention – though it was central to much of the English debate on church-state relations in the 1640s and 1650s. Since Hobbes treated sacraments as purely commemorative ceremonies, which neither conferred nor confirmed grace (Lev 35: 286/221), to bar someone from the Lord's Supper would have little point in his theory. But Hobbes does discuss greater excommunication, by which sinners were publicly declared to be in a state of damnation – or, as St Paul put it, delivered to Satan – and by which the faithful were forbidden to associate or even eat with them (Lev 42: 349–51/276–7; DC17:25–6). Much of what he said was close to orthodox Anglican theory. Once again, however, he extended the theory to prove conclusions which were likely to displease Anglicans, for he deprived the clergy of all their traditional powers.

While Erastus asserted that excommunication was a human invention, without biblical precedent, Hobbes held that it did have scriptural warrant, and equated it with delivery to Satan – which Erastus claimed was an extraordinary Apostolic power. Erastus denied that Christ's words about heathens and publicans imposed any obligation upon Christians to avoid the company of excommunicates. In Hobbes' opinion, the point of the words was precisely to 'forbid them' to associate or even 'eat with a man Excommunicate' (Lev 42: 350/276; DC17:25–6).[46] Like Anglicans, Hobbes argued that the sovereign could never be excommunicated (Lev 42: 352/278; DC17:26), that the purpose of excommunication was to encourage sinners to repent, that the censure therefore only operated upon believers (Lev 42: 350/277), and that excommunication of parents or masters did not terminate the duties towards them of children or servants (Lev 42: 353/279).

He also claimed that the power of judging who should be excommunicated was vested not in the pope as the successor of St Peter, but in the whole church (Lev 42: 348–9/275–6; DC17:25) – a position which was likewise maintained by most opponents of papal pretensions in the later Middle Ages and early modern period. The question which divided these thinkers was: what constitutes the church? One idea, to which the sixteenth-century conciliarists Almain and Mair appealed, was that the primary meaning of the church is the prelates or a General Council.[47] De Dominis, who was much indebted to these authors, and whose views on such matters were largely typical of Anglicans, declared that ecclesiastical jurisdiction had been given 'not to the community but to the prelates', or bishops. In Hobbes's time the common Anglican view was that the bishops possessed the power of excommunication – and therefore *were* the church in the relevant sense. We saw that Hooker and other Anglicans argued that in a Christian commonwealth, church and state are a single body. Hooker claimed that legislative power over both religious and secular affairs resided in the whole commonwealth, or its representative institutions. He explicitly rejected the contention that 'the prelacy' alone possesses power in ecclesiastical matters. Yet when he discussed bishops he asserted that from the earliest times they had held powers of spiritual jurisdiction which were denied to lesser ministers and to the secular authorities.[48] In Anglican theory, bishops possessed the power to excommunicate, and so could be described as the church in a certain sense; but in another sense the church was the whole commonwealth – for ecclesiastical matters were subject to laws made by the state. The church as a law-making institution, then, was identical with the state in a Christian commonwealth; but the church as a repository of spiritual power was the clergy, and in particular the bishops.[49] What Hobbes did was to combine these two meanings of 'church'. He held that there was *no* sense in which bishops or clergy were the church. To say that the church has the power to judge who is worthy of excommunication, he concluded, is to grant this power to the whole community, and in a Christian commonwealth to the sovereign. This position was close to ideas expressed by Marsilius of Padua in the fourteenth century, and by Thomas Cranmer in the sixteenth.

According to Marsilius, the church is the whole body of the faithful in any community. Clerics, he argued, are to specify for what crimes people deserve to be excommunicated, 'but whether the person who is charged with such a crime has committed it, must be judged not by the bishop or priest alone but by the whole body of the faithful in that community'. Hobbes, of course, granted clerics no power even to specify crimes. Early in the English Reformation, Thomas Cranmer, Archbishop of Canterbury, argued that it was the sovereign and his laws that specified who had the right of excommunication. If the law happened to grant this right to bishops or priests, then they could excommunicate, but 'they that be no priests may also excommunicate, if the law allow them thereunto'. Like Hobbes – and unlike the vast majority of later English churchmen – Cranmer also held that election alone (without ordination or consecration) was sufficient to make a man a priest or bishop.[50]

In *De Cive*, Hobbes argued that it was the church which possessed the power to judge whether a particular person is worthy of excommunication. Clerics were simply to pronounce the sentence which the church had decreed. It is sinners, he said, who deserve excommunication, but sin is a breach of law, and to establish whether someone has in fact broken the law it is necessary to consult the interpreter of the law. In a Christian commonwealth this is the church represented by the sovereign. The sovereign, therefore, judges whether the person should be excommunicated or not. Once this judgement has been given, the clergy are obliged to pronounce sentence accordingly (DC17:25). In *Leviathan*, Hobbes added a few minor points to the theory, but did not substantially change it. He observed that since excommunication could achieve its purpose only in the case of believers, it was pointless to excommunicate non-believers. In the early church, he argued, excommunication 'was used onely for a correction of manners, not of errours in opinion' (Lev 42: 350/277). But the rule of manners – that is to say, external actions – in any commonwealth is the law of the sovereign (Lev 42: 353/279). So 'the Excommunication of a Christian Subject, that obeyeth the laws of his own Soveraign, whether Christian, or Heathen' is not 'of any effect' (Lev 42: 352-3). The same principles are easily deducible from the arguments of *De Cive* (DC17:25).

De Cive's position on excommunication is essentially the same as that of *Leviathan*. In both works, Hobbes' teachings are irreconcilable with orthodox Anglicanism. Plainly, Hobbes was well informed about contemporary debates on church-state relations, particularly between Anglicans and Catholics. On many points, he adopted Anglican principles; but he adapted and extended those principles to subvert the doctrines of the Anglicans themselves. It is sometimes suggested that Hobbes' religious outlook was close to that of William Chillingworth and other members of the Tew Circle. Chillingworth, however, defended divine right episcopacy, while Hobbes attacked it.[51]. As we shall see in Chapter 6, the two men did share a number of opinions, and both advocated a large measure of religious toleration. But there were also major differences between their religious principles.

6
God, Religion and Toleration

Throughout the seventeenth century much ink was spilled in debating the question of church-state relations in England. Hobbes contributed to the discussion in *Leviathan* and elsewhere. A problem which was closely associated with church-state relations was that of toleration – and this topic also aroused heated discussion, not least in the 1640s. Independents made common cause with Erastians and with the sects to prevent the institution in England of a rigid Presbyterian system on the Scottish model. But the Independents and their allies were themselves divided on just what degree of toleration should be introduced. A number of issues underlay the debate. One was the extent to which the civil magistrate possessed the authority to interfere in religious affairs. Another was the question of whether individuals have a right to worship God according to the dictates of their consciences. A third was the problem of the nature of heresy and of what if any punishment it merited. Finally, some writers (including Chillingworth and Falkland) argued that there were very few essential Christian doctrines and that it would be wrong to persecute anyone for maintaining (or failing to maintain) a religious belief which the bible neither condemned nor prescribed.

Hobbes said a great deal that was relevant to all these issues. As we have repeatedly seen, his thinking on other points was often close to that of Anglican royalists. It is sometimes suggested that his religious doctrines were much the same as those of Falkland and Chillingworth, both of whom sided with the royalists largely for religious reasons (and especially through fear of puritan intolerance), and both of whom died in the Civil War – Falkland on the battle-field, and Chillingworth in a parliamentary gaol. Of course, Hobbes held that the civil magistrate had the authority to regulate the religious as well as the civil affairs of his subjects. This was a position commonly taken by those who argued against toleration. But he shared other principles with those who advocated toleration, and like Chillingworth he reduced the essential doctrines of Christianity to a minimum. Hobbes argued for a minimalist religion (that is to say, one which consisted of a very small number of fundamental beliefs), and for freedom of enquiry in all areas where reason was capable of adding to human knowledge. At the same time he wished to prevent the spread of ideas which threatened to undermine state security. In Hobbes' view, churchmen were the group most responsible for disseminating seditious opinions, and he claimed that they had corrupted the Christian message to serve their own ambitions by mixing it with the false notions of Greek philosophers. The last part of *Leviathan* is largely taken up with invective against such clerics, and Hobbes frequently reverted to the same theme elsewhere. *Leviathan* is a philosophical treatise on politics. It is also a polemical attack on the clergy and their dangerous opinions, and an impassioned call for educational reform; 1651 was a particularly propitious date to publish on such questions, for the schemes of the Presbyterians had by then been frustrated, and ecclesiastical authority had largely been destroyed. As Hobbes put it, there was then 'no church in England' (*Answer to Bramhall* in EW4:355).

If Hobbes' religious ideas were basically the same as those of Falkland and his friends (the 'Tew circle') it is difficult to explain why *Leviathan* attracted such bitter criticism from Anglicans – including Falkland's admirer and defender Henry Hammond. Certainly some of Hobbes' religious teachings were close to those of the Tew writers. Like them, he argued that according to the bible few beliefs are necessary to salvation, and like them he advocated toleration of people who expressed harmless though unorthodox religious opinions. But on other points he differed drastically from them. They were Arminians who held that God will save everyone who does not wickedly reject Him. Hobbes, in contrast, said that the bible tells us that God will save only those to whom he has arbitrarily decided to grant Christian belief. They broadly supported the Anglican church as it had existed before 1640, and they defended episcopacy against puritan attack. In *Leviathan* Hobbes welcomed the destruction of episcopacy and suggested reforms which would transform the church and the universities. Finally and very importantly, the Tew writers – like most Christians – believed that Scripture contains supernatural truths which God has revealed to man. It is doubtful that Hobbes agreed. All these points will be discussed below, but it makes sense to begin with what Hobbes thought about God.

1: GOD

Hobbes notoriously attracted charges of atheism from his contemporaries. When Roger Coke came to survey the section on religion in *De Cive*, he soon concluded that the task was a waste of time. 'It is not worth the examining, what he would have under the title of *Religion*', said Coke, 'for men say, the man is of none himself, and complains (they say) he cannot walk the streets, but the Boys point at him, saying, There goes HOBBS the Atheist!' Plainly, if Hobbes *was* an atheist, this fact is likely to colour how seriously we take what he had to say on religious matters. At one time, it was widely accepted that Hobbes was indeed an atheist; but more recently he has increasingly come to be seen as a theist and a Christian, though of a rather unorthodox kind. As we shall see there are no good grounds for supposing that Hobbes disbelieved in God (at least in some sense of the term God). Strictly he was no atheist, but his views on God and on revelation were so different from those of most of his contemporaries that they felt justified in calling him an atheist – a word which they often used rather loosely.[1]

Hobbes sometimes argued that we can trace back effects to their causes in a constant chain until 'we come to an eternal, that is to say, to the first power of all powers, and first cause of all causes. And this is it which all men call by the name of God' (EL 1:11:2).[2] The same idea that God is 'the eternal cause of all things' (*Of liberty and necessity*, EW4:246) featured in the case which he mounted for determinism during his controversy with Bramhall. Occasionally Hobbes supported the notion that there is a God in other ways. 'It is very hard to believe', he claimed, 'that to produce male and female, and all that belongs thereto, as also the several and curious organs of sense and memory, could be the work of anything that had not understanding' (*Decameron physiologicum*, EW7:69–180, at 176; cf. *De Homine* 1:4 in LW2:6; 12:5 in ibid., 107). But it was to the argument from causation that he most commonly appealed in order to show that God exists. Variants of the argument were voiced by Aristotle and Aquinas, and Hobbes' contemporaries commonly gave it prominence amongst the proofs of God's existence. Grotius, for instance, claimed that there had to be a first cause, and that this was what the name God signified. 'Since nothing occurs without a cause', said Lord Herbert of Cherbury, 'we must reach some First Cause in the last analysis'.[3]

Hobbes' use of the argument was notable only for its lack of rigour. He never presented it with any great degree of precision. Moreover, he sometimes came close to contradicting it. The argument purported to show that there must be an eternal, unmoved mover, namely God. But in *De Corpore* (chapter 26; EW1:412; LW1:336) he suggested that an infinite regress of moved movers was possible. Still more striking is a passage in his reply to Thomas White. There Hobbes roundly asserted that no proof of God's existence has yet been found, and that it is unphilosophical to attempt to find one. Some, he said, 'have declared that they have demonstrated the existence of God, the Creation of the world, and the immortality of the human soul'. But they 'could not show' that these things were true (AW26:6; f.288v). Those 'who declare that they will show that God exists', he asserted, 'act unphilosophically' (AW26:2; f.287).[4]

So Hobbes frequently asserted that God does exist, and sometimes implied or stated that this was provable. But on other occasions he undermined a principal argument in favour of this thesis, or denied that God's existence is demonstrable. There is no reason to think that he disbelieved in God. His inconsistencies on the problem of whether the deity's existence may be proved do strongly suggest, however, that his political system was in no important respect dependent on God. For his system was intended to be one in which each conclusion was rigorously proved, but he showed no great concern with demonstrating that God existed.

Hobbes said that there is a God, but added that we can know little of His nature. 'And thus all that will consider, may naturally know *that* God is, though not *what* he is' (EL 1:11:2; cf. Lev 34: 271/208; AW 35:16; f.396–396v). The claim that reason alone is insufficient to tell us much about God or true religion was commonplace. According to Aquinas, God is 'simple form', but 'reason cannot reach up to simple form so as to know what it is; but it can know whether it is'. Others said much the same thing.[5] There were, indeed, differences of opinion on the exact extent of the religious information that reason can provide. Ockham and Calvin, like Hobbes, stressed the blindness of reason in matters of faith, while Aquinas and especially Grotius took a more optimistic view of its abilities. Aquinas argued that even if we cannot know precisely what God is, we can at least know some of the things that He is not. He claimed, for instance, that reason shows God is not corporeal, since He is the first mover, but 'no body is in motion unless it has been put in motion'. Hobbes held that God *is* body, and as we have seen he expressed doubts about the argument from causation to an unmoved first cause or first mover. But like Aquinas, Hobbes admitted that natural reason warrants us in using 'Negative Attributes' of God (Lev 31: 251/191). A number of thinkers held that this 'negative way' of knowing God was the only one available to reason. St Thomas himself, however, argued that we can make positive assertions about God, attributing to Him perfections which we perceive in the creatures He has made. Since these perfections 'flow from Him to creatures', argued Aquinas, they must exist 'in God in a more eminent way than in creatures'. Hobbes admitted that reason justifies us in using 'Superlatives, as *Most High, most Great*, and the like' of God (Lev 31: 251/191). We do this not because we truly understand God's nature, he argued, but because it makes sense for us to honour an omnipotent being (Lev 31: 246–51/187–91). This implies that we can know that God is omnipotent, and therefore that his nature is not *wholly* incomprehensible – despite what Hobbes says elsewhere (Lev 34: 271/208).[6]

Though Hobbes sometimes stated or implied that God's existence was provable, he never provided a satisfactory proof and it is doubtful that he thought one was available. Yet he wrote at some length (DC 15; Lev 31) on the 'naturall Duties' or 'Praecepts' which 'are dictated to men, by their Naturall Reason onely' 'touching the Honour and Worship of the Divine Majesty' (Lev 31: 248/188). The problem here is how we can have duties towards a being of whose existence and attributes we cannot be certain. Of course, in Hobbes' theory, if someone happens to believe in God, and to believe that God will inflict eternal punishment upon him unless he perform a certain act, then he will perform that act; but this does not show that he is obliged to perform it. People foolishly act through fear of such imagined powers as ghosts or disembodied spirits, though they have no obligation to do so, and though ghosts do not in fact exist. Again, some obey the pope against the commands of their civil sovereigns, since they believe that he can inflict great punishments upon them; but he cannot in fact do so, and they are under no obligation to obey him. Mere belief creates no obligations. How, then, can those who believe in God come to have any genuine obligations towards him?

Hobbes sometimes suggested that God's existence cannot or has not been demonstrated, but he nowhere implied that it can be disproved – as, for example, the pope's claims to power can be. He acknowledged that people commonly come to believe that there is a God, either because they desire to know causes and are led by searching into them to believe in a first cause, which is God (Lev 12: 77/53; 11: 74/51), or because through ignorance they fear invisible powers (Lev 11: 75/51; 12: 76/52; 6: 42/26). God, in his view, is a device used by men to account for their good or evil fortunes and to provide explanations for events which they cannot otherwise explain. Though people might be led to believe in God for bad reasons – for instance, through ignorance – there were also good reasons which pointed in the direction of a deity, though they did not perhaps formally demonstrate it. '[I]t is impossible', he said, 'to make any profound enquiry into naturall causes, without being enclined thereby to believe there is one God Eternall' (Lev 11: 74/51).[7] The notion that God exists, then, is compatible with reason and indeed suggested by it, while reason shows that there are no incorporeal spirits and that the pope has no jurisdiction over civil sovereigns.

Like Aquinas, Hobbes argued that not all people are necessarily led by reason to acknowledge that there is a God, for they are often too busy or too inept at reasoning to follow the proofs of His existence (DC 14:19 Annot)[8] – and, as we have seen, Hobbes (though not Aquinas) had doubts that there were any strict proofs. But plainly many acknowledged that there is a God, and Hobbes affirmed that these people have duties towards Him. Though he maintained that we can know nothing about what God is, he did suggest that those who acknowledge His existence will also believe that He possesses irresistible power. It was in virtue of this power, said Hobbes, that He holds sovereignty over people who think that He exists and also that He concerns Himself with the actions of mankind (Lev 31: 246/187). Natural reason, he claimed, taught us what sort of honour is 'naturally due to our Divine Soveraign' (Lev 31: 248/188).

According to Hobbes, reason dictates that we honour God by our inward thoughts, and also by outward actions. Some actions, he said, 'are Naturally signes of Honour', and these could never be separated from the worship of God. He instanced 'decent, modest, humble Behaviour'. Again, some acts were always signs of dishonour, and could never 'be made by humane power a part of Divine worship'. But there were an infinite number of Actions, and Gestures, of an indifferent nature' (Lev 31: 253/192). Whether one of these actions counted as a sign of honour in a particular commonwealth depended on public conventions, and these were to be defined by the sovereign, who was therefore empowered to regulate the rites and ceremonies which constituted worship. Hobbes argued that 'Reason directeth not onely to worship God in Secret; but also, and especially, in Publique'. The essence of public worship, he insisted, was uniformity, and he spelled out that 'where many sorts of Worship be allowed . . . it cannot be said there is any Publique Worship, nor that the Commonwealth is of any Religion at all' (Lev 31: 252-3/192). In addition to setting out some basic rules on worship, he argued, reason also persuaded those who believed in God that He was the author of human nature, and that the laws of nature were His laws (Lev 31: 248/188; EL 1:17:12). For believers, therefore, the duty to obey natural law – and also the duty to obey civil law, since that duty was itself derived from the laws of nature – was underpinned by the will of God, who would enforce obedience with natural sanctions (for instance by punishing intemperance with disease) (Lev 31: 253-4/193).

There was nothing particularly novel in Hobbes' contention that reason alone can tell us little about religious matters. On the contents of natural religion, he differed from a number of Catholics who argued that reason itself demonstrated the need for a priesthood which held power over the laity – though only grace showed that this priesthood was the Roman Catholic clergy.[9] Again, his claims that God is material and that religion is often rooted in fear and ignorance were unorthodox. But much of what he had to say about the limitations of reason in religious matters was commonplace enough. It was his attitude towards faith and revelation which led to the charges of atheism which were levelled against him.

2: FAITH, SALVATION, REVELATION, PROPHECY AND MIRACLES

Hobbes' contemporaries held that reason could supply people with only limited information on religious truths. What was needed for a full account of such verities, they said, was revelation from God Himself. Fortunately, they held, such revelation had in fact taken place, and was recorded in Scripture and perhaps in the traditions of the church. It was by faith, they proceeded, that people came to acknowledge the supernatural truths set down in the bible. Aquinas contrasted the demonstrative knowledge available through reason with the insights which faith might attain, insisting that supernatural truths could not be demonstratively known. Calvin held that reason was practically useless in religious matters, but he claimed that faith could confer firm and sure knowledge of God's revealed will. Aquinas had similarly argued that the Holy Ghost grants believers the gift of knowledge, by which they come to have a sure and right judgement of the things of faith. Though this knowledge was intuitive and not demonstrative, it was knowledge none the less. Chillingworth likewise held that once the historical evidence on the subject has induced people to believe the claims of Scripture, the Spirit of God will give them knowledge of 'what they did but believe'. There was wide agreement, then, that people can attain *knowledge* on matters of faith, though this knowledge was not acquired through rational demonstration.[10]

Of course, there were differences of opinion on the relationship between reason and faith. Calvinists commonly stressed the fallibility of human reason not only on religious questions but also upon more mundane matters. Chillingworth, Falkland and Grotius, by contrast, emphasised the utility of reason, even on theological questions – and this was why they attracted charges of Socinianism.[11] Grotius' *De veritate religionis Christianae* was a disquisition on the reasonableness of Christianity. 'For my part', declared Falkland, 'I profess myself not only an anti-Trinitarian, but a Turk, whensoever more reason appears to me for that, than for the contrary'. Chillingworth, said his Calvinist critic Cheynell, 'did runne mad with reason, and so lost his reason and his religion both at once'. Grotius and his followers argued that reason operating upon historical evidence was sufficient to show that Christianity is superior to other religions. Reason could not demonstrate Christian doctrines, but it supplied convincing grounds for belief in their truth. '[S]carce anything', argued Falkland, 'is in nature capable' of 'absolute Demonstrations' 'but lines and numbers'. Reason, he held, provided as much support for Christian teachings as for any other truths except those of such demonstrative systems as geometry and mathematics.[12]

Hobbes, of course, held that reason could give us demonstrative truths not only on geometry but also on ethical and political questions. In his view, the achievement of *De Cive* was precisely that it set forth a science of politics. He emphasised the reasonableness of his own political theory, but said little to suggest that Christianity was reasonable. Faith, he claimed, was brought about by teaching and similar means, though not all who were taught came to believe – and '[i]t is certain therefore that Faith is the gift of God, and hee giveth it to whom he will' (Lev 43: 406/324). Hobbes argued that according to Scripture, salvation requires faith as well as obedience (Lev 43: 403/321). Whether we have faith or not depends upon whether God has chosen to grant it to us. Hobbes' God is the arbitrary Lord of the Calvinist tradition – though unlike Hobbes, Calvin held that believers *know* truths revealed by faith. 'Faith', said Calvin, 'consists in the knowledge of God and Christ'.[13] Hobbes, by contrast, insisted that we can have no knowledge of the things of faith (Lev 43: 405–6/323–4).

Hobbes did, indeed, argue that beliefs can sometimes be as securely based as knowledge: 'Belief, which is the admitting of propositions upon trust, in many cases is no less free from doubt, than perfect and manifest knowledge'. Much of what was recorded in history books, for example, could readily be accepted as true. Where many independent witnesses, with no reason to deceive, had reported some past event there was little ground for doubt that it had happened (EL 1:6:9). As we have seen, Hobbes held that Scripture gave a reliable account of the various human actions which it chronicled (see Chapter 5, section 1 above). But his theory left it unclear that there were any good grounds for believing that the *supernatural* events recorded in the bible were authentic. Someone may indeed tell me that God has spoken to him in a dream or vision; but to 'say he hath spoken to him in a Dream, is no more than to say he hath dreamt that God spake to him; which is not of force to win beleef from any man, that knows dreams are for the most part naturall'. Visions, he proceeded, are simply dreams 'between sleeping and waking'; so there was good reason for doubting the claims of anyone to have had an immediate revelation from God, or to be a true prophet (Lev 32: 257/196). As his critic William Lucy put it, the effect of Hobbes' doctrine on this point was 'to render Christian religion suspected'.[14]

Hobbes examined the Scriptural message on what constituted a true prophet, concluding that such a person might be recognised by two marks – '[o]ne is the doing of miracles; the other is the not teaching any other Religion than that which is already established' (Lev 32: 257/197; DC 16:11). He claimed that the age of prophets and miracles was now over, and this line was common enough amongst Protestants (Lev 32: 259/198).[15] Some puritans in the civil war period argued that God was still revealing new religious truths to people – or giving them 'new light', as the expression went. But more conservative thinkers held that the bible provided a complete and final account of Christian revelation, and condemned as fanatics or enthusiasts those who claimed that God had shown them new truths.[16]

On two important points Hobbes differed from these conservatives – and the differences go a long way towards explaining why people doubted how seriously he took revelation. Firstly, by claiming that Scripture requires true prophets to teach only the established religion, Hobbes made revelation dependent on the will of the sovereign, for it was the sovereign who established religion. 'Every man', said Hobbes, 'ought to consider who is the Soveraign Prophet; that is to say, who it is, that is God's Vicegerent on Earth; and hath next under God, the Authority of Governing Christian men; and to observe for a Rule, that Doctrine, which in the name of God, hee hath commanded to be taught'. If the teachings of any self-proclaimed prophet were contrary to this doctrine, then they were to be rejected (Lev 36: 299/232). God therefore reveals religious truths to us only through the agency of our sovereign or of those who act with his approval. Hobbes found this thesis in Scripture, but confirmed it by rational means, arguing that the claim to be a prophet was in effect a claim to govern – '[f]or he that pretends to teach men the way of so great felicity, pretends to govern them' (Lev 36: 297/230). To accept such a claim, he said, was a recipe for civil war, and he instanced the case of people who were bewitched into rebellion by their fellow subjects, though the latter had no 'other miracle to confirm their calling, then sometimes an extraordinary successe, and Impunity' (Lev 36: 300/232). This was a clear reference to Cromwell and his army, whose propagandists commonly asserted that their successes testified to their special providential role.[17]

Secondly, Hobbes' discussion of miracles was shot through with scepticism. He did indeed argue that ability to perform miracles was one of the signs by which true prophets had been marked out in biblical times. But he stressed the 'ignorance, and aptitude to error generally of all men, but especially of them that have not much knowledge of naturall causes, and of the nature and interests of men'; such people were liable 'by innumerable and easie tricks to

God, Religion and Toleration

be abused' (Lev 37: 304/236). In view of 'this aptitude of mankind, to give too hasty beleefe to pretended Miracles', Hobbes recommended extreme caution on the subject, and once again stressed that the performance of miracles was no sufficient mark of a true prophet unless he also taught doctrine approved by the sovereign (Lev 37: 303/237). Protestants and humanists scoffed at medieval monkish legends of miracles, but accepted the literal truth of those recorded in the bible. It is far less clear that Hobbes did so.

His account of revelation could easily be read as implying that belief in the Christian message is foolish. His discussion of what that message is was also highly unorthodox. Hobbes denied that Moses wrote the books of the Pentateuch as we now have them – a viewpoint which was only very rarely adopted by earlier authors (Lev 33: 261–2/200).[18] He rejected the common notion that Scripture says the damned will suffer eternally in hell, claiming instead that it asserts they will be resurrected here on earth and will then die a second and final death (Lev 38: 315/244–5; he changed his mind on the details of what will happen to the reprobate: cf. Lev 44: 432–3/345–6, the Latin version in LW3:466–7, and the *Answer to Bishop Bramhall's book* in EW4:359). This, again, was most unusual, though it is true that a number of English thinkers in the seventeenth century came to argue against traditional ideas of hell.[19] Applying his notions on representation to the question of the Trinity, Hobbes affirmed in *Leviathan* that God's person had been borne three times – firstly by Moses and the High Priests, then by Christ, and finally by the Apostles and their successors (Lev 16: 114/82; 42: 339–41/267–9). Later, he retracted this opinion, ostensibly because John Cosin (who became Bishop of Durham at the Restoration) persuaded him that it was not warranted by Scripture (*Answer to Bishop Bramhall's book*, EW4:117; appendix to the Latin *Leviathan*, chapter 3, LW3:563).

Hobbes' arguments on the Trinity, Hell and other questions served to confirm his general thesis that the clergy had misinterpreted the bible. The arguments irritated clerics and doubtless amused Hobbes. They did little to support his wider theories, however. For instance, nothing could be concluded about civil or ecclesiastical power from the fact that Moses was (or was not) a member of the Trinity. Moreover, on Hobbes' own account such questions as the nature of the Trinity were of no importance to salvation – for the '[o]nely Article of Faith, which the Scripture maketh simply Necessary to Salvation, is this, that JESUS IS THE CHRIST' (Lev 43: 407/324; DC 17:7).

In a book published in 1617 the Jesuit John Sweet attacked the church of England for tolerating people of widely varying religious views. 'I understand it is an opinion growing into fashion among you', he said, 'that a man may be saved in any Religion, so he believe in Christ'. Sweet cited a treatise printed in 1596 and written by Thomas Morton of Berwick. In this work Morton argued that 'the foundation of the Gospell . . . is faith in Jesus Christ the sonne of God and the saviour of the world'. He claimed that people who had faith in Christ would be saved even if through ignorance they also believed doctrines which were in fact incompatible with this fundamental truth. Much the same minimalist view of Christian fundamentals was later given more famous expression by Hales, Falkland and Chillingworth.[20]

Aquinas and other writers had distinguished between doctrines which required *explicit* belief and those in which *implicit* belief was sufficient. According to Aquinas, salvation was impossible for anyone who did not believe in the incarnation of Christ and in the Trinity. There were a large number of other true doctrines, he held, but not everyone was well enough informed to be aware of them. It was sufficient for salvation that you have an *implicit* faith in such points; that is to say, you have to believe that what Scripture says is true, and therefore implicitly believe all that it in fact teaches. But you do not have to believe any except the fundamental doctrines *explicitly* until it becomes clear to you that the bible does indeed teach them. Just what you need to believe, then, depends on your own particular circumstances and education. A small number of beliefs are always required, and a much larger number may become so. What Chillingworth and like-minded authors did was to reduce the number in both categories, with the idea of making the pathway to heaven as easy as possible.[21]

Where 'little is given', said Chillingworth, 'little shall be required'. If God, he argued, has granted us means only to believe 'that God is and that he is a rewarder of them that seek him' then this will be sufficient in the end to bring us to Christ and so to salvation. 'Whosoever dies with faith in Christ', he claimed, 'and contrition for all sins' will be saved. Some at least could win salvation by believing very little indeed, though more might be required from people of greater intelligence or educational attainments. But Chillingworth was always very hesitant to say just what it was that even the best educated had to believe. As his friend Clarendon put it, he developed such a 'Habit of doubting, that by Degrees He grew confident of nothing, and a Sceptick at least, in the greatest Mysteries of Faith'.[22]

John Hales argued that many things in Scripture are 'seemingly confused', 'which brings infinite obscurity to the text'. He therefore recommended that '[i]t shall well befit our Christian modesty to participate somewhat of the Sceptic'. Falkland adopted much the same position, and favourably cited the French sceptic Charron. Influenced by authors such as Erasmus, Acontius, Cassander and Grotius, these thinkers of the Tew circle stressed the difficulty of attaining certainty on more than a few points of Christian faith, and condemned persecution of people who held heterodox views. God, they said, is merciful and requires of us only that we do our best to seek Him and obey His will. To assert that people 'who do their endeavours' will none the less be damned, claimed Falkland, 'is to lay their damnation to God's charge'. Someone who held an erroneous opinion through no fault of his own would not be punished for it by God. '[I]f his error be not voluntary', said Jeremy Taylor, 'and part of an ill life, then because he lives a good life, he is a good man', and will go to heaven. Only those who were led to false beliefs by avoidable vices – such as pride or 'stupid carelessnesse' – will be damned.[23]

It is sometimes said that Hobbes' doctrine of salvation was essentially the same as that held by members of the Tew Circle. He did share some of their views.[24] Like Chillingworth and his group, Hobbes argued that few beliefs were necessary to salvation, claiming that the only indispensable article of faith was 'that JESUS IS THE CHRIST' (Lev 43: 407/324). Again, Chillingworth, Hales, Falkland and Taylor adopted a rather dismissive attitude towards the Fathers and traditions of the church. In linguistic skill, said Hales, the Scriptural 'interpreters of our own times . . . have generally surpast the best of the antients'. 'Antiquity', he argued, is 'but man's authority born some ages before us', and like all human authority it is fallible. '[T]he circumstance therefore of time', he concluded, 'in respect of truth and error, is merely impertinent'. Falkland claimed that in 'these more learned times' men 'argue . . . every way more rationally than the ancient Doctors used to do'. Hobbes adopted much the same position. He reverenced those ancient writers 'that either have written Truth perspicuously, or set us in a better way to find it out our selves'. But 'to the Antiquity itself', he trenchantly added, 'I think nothing due' (Lev 'Review and Conclusion', 490/395).[25]

In other respects, however, Hobbes' teaching was very different from that of these thinkers. They adopted the conventional view that nature and Christianity prescribed many actions which had to be performed even if this meant disobedience to the sovereign. Hobbes, by contrast, argued that Christianity confirmed the subject's duty of absolute obedience to the ruler. They were drawn to emphasise the small number of fundamental doctrines by their belief in God's charity. God, they said, damned only those who voluntarily rejected him. For Hobbes, however, salvation and damnation were consequences of God's arbitrary decree. God was the first cause, and all things inevitably followed from that cause, regardless of whether they happened to correspond to human notions of fairness. God's sovereignty was derived from his omnipotence, which gave him the right to afflict even such a righteous man as Job (DC 15:5–6; Lev 31: 246–7/187–8).

Like Falkland (and the rest), Hobbes argued that the usual reason why people believed the Christian message in Christian countries was that there they had been taught it by their parents and pastors; elsewhere people did not believe it since they had not been taught it (Lev 43: 406/324). Falkland concluded that God will not save or damn us for simply believing what we have been taught: 'I cannot see why he should be saved, because by reason of his parents' beleef, or the Religion of the Country, or some such accident, the Truth was offered to his understanding, when, had the contrary been offered, he would have received that'.[26] Hobbes admitted that what people believe depends largely upon such accidents as where they happen to live, but none the less argued that according to the bible salvation absolutely requires belief that Jesus is the Christ. Falkland was an Arminian, who stressed the role of free will in the scheme of salvation, while Hobbes was a predestinarian who wholly rejected the idea that the will is free.

Chillingworth and his friends adopted a minimalist account of Christian fundamentals in order to argue that all who tried to do so could get to heaven. Hobbes held that faith is the gift of God, and that those to whom He fails to grant it can never fulfil the Scriptural criteria for salvation, however hard they try. The minimalism of the Tew writers flowed from their conception of God as a merciful father. Hobbes' minimalism, on the other hand, was intended to establish two points about the political implications of Christianity. The first was that Christian doctrine could never warrant disobedience to the sovereign. Scripture itself, said Hobbes, made obeying the sovereign a condition of salvation (Lev 43: 404/322). It also specified that to be saved a person had to believe 'this Foundation, *Jesus is the Christ*' and 'all that hee seeth rightly deduced from it' (Lev 43: 411–12/328). Christians regarded the bible as the word of God. The bible said that to be saved you had to obey your sovereign and believe that Jesus was the Christ; but someone who did these things would in no way endanger the state. Hobbes argued that if Christians followed the bible's precepts there would be little need for them ever to face martyrdom at the hands of some persecuting pagan prince. For 'what Infidel King is so unreasonable' as to persecute a subject 'that waiteth for the second comming of Christ, after the present world shall bee burnt, and intendeth then to obey him (which is the intent of beleeving that Jesus is the Christ,) and in the mean time thinketh himself bound to obey the Laws of that Infidel King . . . ?' (Lev 43: 414/331). Since Christianity encouraged obedience and not sedition, the infidel would tolerate it.

The second point of Hobbes' discussion of salvation was to promote intellectual freedom and tolerance. If the bible required belief in only a few doctrines then it made sense for the Christian magistrate to provide for the teaching of those points but to allow freedom of inquiry in other areas. Like Falkland and his friends, Hobbes used the idea that there were few fundamental Christian beliefs to argue in favour of toleration.

3: TOLERATION AND CONSCIENCE[27]

Hobbes gave the sovereign full power to make laws governing people's actions. A ruler, he held, had a duty to use this power in order to promote the public good. So sovereigns should prohibit the preaching of doctrines which militate against the public interest, and promote the teaching of socially useful precepts – such as the rule requiring that subjects obey their ruler. Someone was needed to judge which doctrines are in fact socially useful and ought to be taught. If this judge was anyone other than the sovereign, then sovereignty would be divided and therefore destroyed. So the sovereign alone was to decide what doctrines should be taught (Lev 18: 124–5/91). Hobbes vigorously rejected all claims that individual subjects have a right of free speech, or a right to act in accordance with their private consciences. Such claims, he held, threatened to ruin the state. This might seem to suggest that Hobbes was an advocate of persecution, but as we shall see that is only very partially true. He did indeed reject some of the arguments for toleration put forward by radical puritans, but he accepted much of the liberal Anglican case mounted by men like Falkland and Taylor.

The notion that *'whatsoever a man does against his Conscience, is Sinne'*, argued Hobbes, is a 'doctrine repugnant to Civill Society' (Lev 29: 223/168; cf. DC 12:2). He inveighed against people who claimed that God had directly inspired them, and who took this inspiration rather than the law of their sovereign as the rule of their actions (Lev 29: 223–4/169; DC 12:6). In the 1640s some puritans argued for religious toleration on the grounds that we have a duty to follow the dictates of conscience even if they lead us into error. According to the radical Independent John Goodwin 'a man is not bound in conscience to doe any thing that is commanded' as long as 'his judgement and conscience remain considerably doubtfull, and unsatisfied touching the lawfulnesse of it'. He spelled out that this applied even 'though both the authority whereby it is commanded, yea and the thing it selfe which is commanded, be never so lawfull'. It was the individual's duty, he argued, to enter into 'a serious and conscientious debate' on the lawfulness of anything decreed by authority, and to obey the decree only if he was persuaded that it was indeed lawful. '[T]his is a mans duty that God requires', said the Independent Henry Robinson, 'that conscience should give warrant and direction for every act', for 'man can do nothing religiously without the perswasion of his conscience'. He combined this idea with the belief that God was still revealing new religious truths to the faithful: 'God reveales more and more of the Gospel each day in a fuller and clearer manner'.[28]

The greatest of all liberties, said the poet John Milton, was 'the liberty to know, to utter, and to argue freely according to conscience'. He too claimed that 'new light' had recently 'sprung up' and was 'yet springing daily', at least in London. To force people to act against their consciences, he said, was to make them sin, and that was itself sinful. Writing to Sir Samuel Luke in 1645, the distinguished Parliamentarian soldier Charles Fleetwood argued that 'in these times, wherein we expect light from God, our duty is not to force men but to be tender of such as walk conscientiously'. For even if a man expressed opinions which seemed very unorthodox, it was possible that God had just revealed them to him. According to Goodwin, Milton and like-minded authors, individual rights of conscience were enforceable against the civil magistrate, and it was to defend such rights (amongst others) that parliament had taken up arms against Charles I (or his evil advisers) in 1642.[29]

The obvious problem with the idea that we must always obey our consciences is that conscience could dictate that we do strange and anti-social things. Hobbes was not alone in noting this. The more conservative parliamentarians agreed with men like Goodwin that one reason for fighting against the king's supporters in 1642 had been to vindicate liberty of conscience. But by this term they meant not the freedom of each individual to follow his own inclinations in religious matters, but freedom to worship according to the prescriptions of the bible. The king's ecclesiastical advisers, they argued, had enforced the use of ceremonies which had no scriptural basis and were purely human inventions. To employ such non-biblical rites was to commit the sin of 'will-worship', treating a human decree as though it were divine. Freedom of conscience, said conservative puritans, would be established when pure doctrine was preached and when the scriptural injunctions on worship were obeyed. To suppose that we should obey conscience even against the word of God, said the leading Scots Presbyterian Samuel Rutherford, was to 'make a Pope, and a God of our own conscience'. If someone had an 'erroneous conscience' which misinformed him about his obligations then it was his duty to lay it aside. He vigorously rejected the notion that God was still adding to what had been revealed in the bible.[30]

According to Aquinas, we should follow the dictates of an erroneous conscience, but we sin if we have such a conscience through our own negligence or culpable ignorance. The Elizabethan Jesuit Robert Parsons claimed that 'an erroneous conscience bindeth', and used the proposition to argue that Catholics should be tolerated in England. But he admitted that 'a man may be bound to reforme or alter his conscience', and thought that if Catholics returned to power in England, the Protestants there should be given only a limited time to conform. For most Catholics, the church acting through the pope pronounced infallibly on matters of faith, and it was the duty of individuals to accept its pronouncements.[31]

In the opinion of many Protestants, Scripture contained clear information on religious questions, and God had instituted churchmen as its authoritative though not infallible interpreters. The definitive rules for conduct and belief, said Rutherford, were 'the Law of nature' and 'the word of God', but not 'the erroneous conscience' of 'libertines' which they used to excuse their immoralities. The word of God, said the leading New England Independent John Cotton, was so clear 'in fundamental and principal points of Doctrine' that no one could fail to understand it. If someone who had an erroneous conscience on such matters were duly admonished 'once or twice' he would inevitably see the light. Anyone who persisted in his error after such admonition would be acting hypocritically and against conscience. He might then justly be persecuted. Cotton, like many others, held that the bible put forward an extensive set of clear doctrines. For Hobbes and for such advocates of toleration as Goodwin and Chillingworth only a few dogmas were clearly contained in Scripture.[32]

People who held that God was still revealing new truths could evidently not accept that any existing rule – say, the decrees of a church, or even Scripture itself – was a complete guide to faith. So it made sense for them to stress the rights and importance of the individual conscience, into which (they argued) God might infuse new light. But if conscience could justify unorthodox religious opinions, why could it not also excuse breaches of natural or human law? Pleas for tender consciences, said Rutherford, vindicated antinomians who argued that Christ has freed the saints from the bondage of all laws, human and divine.[33]

Goodwin replied to this type of charge by claiming that some actions can never be performed conscientiously. The natural law was so manifest, he argued, that no one could be ignorant of it. Anyone who broke it would therefore be sinning against his own conscience, whatever he might pretend. The task of the magistrate was to promote the public good, and this was to be done by punishing infractions of natural law. According to the Leveller William Walwyn, the civil authorities ought not to take cognisance of 'matters of opinion' 'any farther, then they break out into some disturbance, or disquiet to the state'. The Independent, said John Cook, 'is a professed enemy to all imperative, co-active violence in matters of conscience, which are not an offence against civill justice'. Others said the same thing.[34]

The idea that the magistrate should not punish people who taught heterodox doctrines was far from universal amongst Independents (Cotton and the leading English Independent Philip Nye, for instance, argued against it). Nor was the attitude confined to congregationalists. Falkland and his friends held that the point of punishment was to uphold the public good, and that this would not be done by inflicting pain on people who maintained harmless though perhaps erroneous opinions. 'It is unnaturall and unreasonable to persecute disagreeing opinions', said Taylor, and he insisted that the purpose of punishment was 'to prevent a future crime', not to combat erroneous opinions. Persecution might lead people to hide their views, but it was unlikely to make them change their minds. According to Falkland, criminals would bring 'destruction to the state' if they were left unchecked. It was the magistrate's duty to 'watch over' the state, but not to concern himself with 'speculative opinions'. He should punish crime – for criminals 'are certaine to hurt others' – but not heresy, for it was unclear that heretics would hurt anyone. By tolerating diverse views, the magistrate would encourage people to make 'a careful search of God's Truth' and so promote a spirit of enquiry which would contribute to the welfare of his subjects. Puritan tolerationists similarly argued that toleration would advance truth and the public good.[35]

Falkland and Taylor – like Goodwin – claimed that there was no point in persecuting people for holding unorthodox opinions. The point of punishment was to discourage anti-social actions, not odd beliefs. Hobbes agreed with all of this. Goodwin, Falkland, Taylor and others put forward a number of reasons in favour of toleration, but underlying their argument was the conviction that we can be certain of only a very few Christian dogmas.[36] Hobbes likewise used doctrinal minimalism to argue for tolerance, inveighing against those who made 'new Articles of Faith, by determining every small controversie', though Scripture had left them undetermined, and thus imposing 'a needlesse burthen of Conscience' upon people (Lev 42: 351/278). Goodwin held that everyone was aware of natural law and could therefore not be excused for breaking it. Similarly, Falkland argued that 'Malefactors plainlie offend against their Consciences', since they must necessarily be aware that their actions are criminal, and Taylor argued that no one can be ignorant of 'his practicall duty'.[37] Once more, Hobbes took the same line: 'Ignorance of the Law of Nature Excuseth no man; because every man that hath attained to the use of Reason, is supposed to know, he ought not to do to another, what he would not have done to himselfe' (Lev 27: 202/152).

Where Goodwin and other puritan tolerationists differed from Taylor and the Tew writers was on the connected questions of whether we should obey the magistrate in doubtful matters, and whether he could bindingly command us to perform religious ceremonies unmentioned in the bible. The Independent Goodwin's talk of conscience was intended to vindicate the individual's right to disobey the civil authorities if he had suspicions that what they commanded was in fact forbidden by God. Like many other puritans, Goodwin argued that the magistrate could make no laws 'in matters of Religion, and which concerne the worship of God'. Scripture alone regulated worship, and where it was silent the individual was free to act as he pleased. On this point, the Presbyterian Rutherford argued similarly. Both 'Jewes and Gentiles', he said, were 'freed in Christ' from the 'Commandements of men' on religious questions. Indeed, he claimed that a ruler's commands in all matters indifferent – that is to say, both religious and civil matters on which the bible and the law of nature did not pronounce – could not bind the consciences of his subjects. Goodwin made the individual conscience the ultimate arbiter of right and wrong. Rutherford rejected this as anarchic, and subjected conscience to the dictates of Scripture as construed by the Presbyterian church. But both absolved the individual from any duty to obey the ruler in doubtful or indifferent matters. By contrast, Anglicans – and Hobbes – stressed this duty.[38]

When Parsons argued that Catholics should be tolerated – since even if they were in error they had a duty to follow their erroneous consciences – Bishop William Barlow replied that people should obey their rulers rather than their erroneous consciences. '[E]ven against a man's Conscience', he said, 'the Prince is to be obeied' unless it could be shown that what conscience dictated was also prescribed by God. 'Obedience', he declared, 'is to be enioined even against Conscience, if it be erroneous or Leaprous'. If the king commands something and I am uncertain whether it is forbidden by God or not then I must obey. Conscience, said Henry Hammond, 'implyes knowledge'; but if what I believe is in fact untrue or if I have doubts, I clearly do not know. If I think there is a possibility that the king's orders are not contrary to God's law I should therefore obey them. It is, said Hammond, 'my duty, and part of my Christian meeknesse, in doubtfull matters to take my resolution from those whom God hath placed over me'. Of course, Hammond and others thought that a number of matters were *not* doubtful since God and nature had imposed quite an extensive set of prescriptions on people. Hobbes himself held that there are cases in which we ought not to obey the sovereign – for instance, if he

God, Religion and Toleration 155

commands us to kill or accuse ourselves. The range of these cases was much narrower in Hobbes' theory than in, say, Hammond's. But both writers endorsed the fundamental principle that except in such cases the subject must obey the sovereign.[39]

Anglicans argued that the civil authorities could legislate in matters indifferent, both in the religious and secular spheres, and that subjects would be bound in conscience to obey the laws.[40] The state might therefore prescribe what ceremonies should be adopted in Christian worship. Taylor said that rulers should not persecute people who expressed unorthodox religious views unless their opinions fomented sedition. But he also asserted that the king may make laws regulating public worship, and that conscientious objectors to these laws should not be tolerated. To permit infractions of such laws would lead, he argued, to 'the totall overthrow of all Discipline'. 'For to say that complying with weake consciences in the very framing of a Law of Discipline, is the way to preserve unity, were all one as to say, To take away all Lawes is the best way to prevent disobedience'. True, some people might be troubled in conscience by having to conform to ceremonies which they (wrongly) believed to be unlawful. But there was no reason why the magistrate should make any special provision for these people, especially since by doing so he would be likely to trouble the consciences of other 'pious and prudent' folk who liked the ceremonies.[41] Exactly the same point was made by John Locke in an early essay in which he argued that the civil magistrate 'may lawfully impose and determine the use of indifferent things in reference to Religious Worship', though later he changed his mind on this vital point.[42]

Hobbes, like Taylor, advocated uniformity in outward religious observance, combined with toleration of opinions that are not seditious. The ruler, both men held, should impose a single form of public worship upon his subjects, and should be obeyed in all indifferent matters, but he should not persecute people who express harmless views. Provided that his subjects obey him, the sovereign ought not to pry into the secrets of their thoughts. The 'power of the Law', said Hobbes, 'is the Rule of Actions onely', and should not be extended 'to the very Thoughts, and Consciences of men, by Examination, and *Inquisition* of what they Hold, notwithstanding the Conformity of their Speech and Actions' (Lev 46: 471/378). Hobbes claimed that God should be worshipped 'with magnificence and cost'. He inveighed against extempore prayers and argued that 'to adorn the place of his worship less than our own houses' was a manifest 'sign of contempt of the divine majesty' (EL 1:11:12). In short, Hobbes opposed puritanical ideas on worship and favoured the more sumptuous rites of the Laudian church. Like Laudians, he expressed doubts that the pope was Antichrist (Lev 42: 382/303–4).[43] On these points, and on the questions of uniformity and indifferency, Hobbes' thinking was indeed typical of the conformist Anglicanism of the 1630s.

So far, we have seen that Hobbes advocated a minimalist religion and a wide measure of toleration for doctrines which in no way undermined the public peace. He hoped for a reformation of English society and institutions in accordance with reason and believed that the suppression of all heterodox doctrines – however true or beneficial they might be – would redound to the public detriment, though it might promote the narrow sectional interests of the clergy. He tried to show that clerics had systematically distorted and disguised the truth in their own interests. He also put forward proposals for reform, especially of education. These themes took up much of the last part of *Leviathan*, and featured again in many of his later works.

4: CHURCH HISTORY AND REFORMATION

In the fourth part of *Leviathan* ('Of the Kingdom of Darkness') Hobbes argued that the simple original message of Christianity had been distorted by the addition of false teachings taken from Greek philosophy (Lev 45: 441/353; 46: 459–62/368–70). He claimed that these Greek ideas and other misguided notions had led to the adoption by Christians of a number of erroneous doctrines. In consequence, the church was 'not yet fully freed of Darknesse' (Lev 44: 418/334). Faulty thinking about the powers of churchmen, and about such things as demons, purgatory, ghosts, and incorporeal spirits had obscured useful truths and spread sedition. The function which these false doctrines served was to promote the interests of churchmen. Indeed, the whole of scholastic philosophy was calculated to advance the power of the clergy by keeping layfolk in ignorance.

The absurd philosophy and fruitless controversies of the clergy, he argued, had furthered the ambitions of the pope and his supporters, and more recently those of Presbyterian ministers: 'The

Authors therefore of this Darknesse in Religion, are the Romane and the Presbyterian clergy' (Lev 47: 476/382). Clerics protected their position by suppressing natural science and 'the Morality of Naturall Reason' (Lev 47: 480/385). They perpetuated their errors by controlling the university curriculum (Lev 46: 462–3). It would be of great benefit, Hobbes claimed, if the sovereign reformed the universities by decreeing that the doctrines of *Leviathan* be taught there (Lev 'Review and Conclusion' 491/395).

Few people agreed with this last suggestion, but other aspects of Hobbes' attitude were widely shared. Like many humanists, the great Dutch scholar Erasmus had inveighed against scholastic theologians in the early sixteenth century, decrying the 'newly-coined expressions and strange-sounding words' which they employed. No one, he held, could possibly understand what they meant with their talk about quiddities, ecceities and so on, unless 'he could see through blackest darkness things which don't exist'.[44] Hobbes similarly remarked 'that the Writings of Schoole-Divines are nothing else for the most part, but insignificant Traines of strange and barbarous words' (Lev 46: 472/379). The confused terminology and thought of the scholastics, he argued, were amongst the most important features of 'the kingdom of darkness'. Erasmus contrasted the simplicity of the apostles' teaching with the tortured complexity of scholastic doctrine.[45] Later, such themes became commonplace in Protestant attacks upon Roman Catholicism – though Protestants themselves adopted many scholastic doctrines. The only light which the schoolmen produced, said the cleric Samuel Shaw, was 'the light of Gloworms, and rotten chips', and Pierre Du Moulin inveighed against the 'Schoole-distinctions' with which Bellarmine 'dazels the eyes'. By blinding those they taught with scholastic dogma – drawn especially from Aristotle – Catholics furthered papal ambitions. 'The popes hierarchie', said the ex-Catholic Thomas Abernethie, 'is built upon Aristotles politicks'.[46]

The same broad themes featured in the writings of those who opposed Catholicism as much because they saw it as an obstacle to the advancement of learning as for religious reasons. Bacon condemned the 'vain matter' of scholastic divinity, claiming that the schoolmen had produced only 'cobwebs of learning, admirable for the fineness of thread and work, but of no substance or profit'. The schools, said Edward Lord Herbert of Cherbury, had suppressed 'common notions' (or self-evident truths) and had become 'so sunk in error that there is hardly any type of argument in which' they 'cannot throw dust in the eyes of their disciples by their clever tricks'. The *Exercitationes paradoxiae* of Hobbes' friend Gassendi was largely a disquisition on the ignorance and obscurantism of the scholastics. Schemes for educational reforms with a practical and scientific emphasis were a keynote of the intellectual efforts of Samuel Hartlib's circle, which included the educationalist Comenius, and which was greatly influenced by Bacon.[47]

Hobbes held that the ancient Roman Catholic church had made converts by taking over pagan religious rites and investing them with a Christian significance (Lev 45; *Historia Ecclesiastica* lines 1311–48, in LW5:384–5). The same idea was commonly voiced in Protestant attacks on popery. So too was the notion that papist priests swayed the superstitious populace by pretending to perform magical acts. Taylor spoke of their 'notorious forging of Miracles, and framing of false and ridiculous Legends'. Amongst other examples he mentioned 'their Devils tricks at Lowdon' – a reference to the famous case of the 'devils of Loudun' who had allegedly possessed some nuns until they were exorcised. Sir Kenelm Digby discussed the affair with Hobbes in 1637. John Gee devoted much of his pamphlet *The foot out of the snare* (1624) to exposing the fraudulent miracles of papists. He drew on Samuel Harsnet's *A declaration of egregious popish impostures* of 1603. Harsnet, who became Archbishop of York in 1629, had also exposed a series of fraudulent puritan exorcisms.[48]

The idea that puritan or, more specifically, Presbyterian clerics had taken over the self-seeking and seditious goals and methods of papist priests featured in *Leviathan* (Lev 44: 427/341; 47: 475–6/382). It was a commonplace of anti-Presbyterian polemic in the 1640s, with roots stretching far further back. Writing to Bishop Edwin Sandys in 1573, the Swiss reformer Rudolph Gualter expressed the fear that the establishment of Presbyterianism would bring with it clerical tyranny and 'the beginning of a new papacy'. Erastus vigorously inveighed against Presbyterian tyranny, comparing it with popery. Those who advocated Presbyterianism, said the cleric Cornelius Burgess in 1645, were 'every where' called 'Persecutours' and 'worse than Bishops'. Presbyterians were widely accused of plotting to set up ecclesiastical tyranny by exploiting lay superstition.[49]

Hobbes' comments on popery and magic, and on Presbyterian clericalism were typical enough of much sober and moderate opinion; but some of his ideas for ecclesiastical and educational reform were disturbingly close to those of radical puritans. He claimed that the church was 'not yet fully freed from Darknesse', and pointed to the fact of religious dissension in order to confirm his thesis that 'we are . . . yet in the Dark' (Lev 44: 418/334). Of course, the light to which he looked to illuminate this darkness was the light of natural reason. Radical puritans, by contrast, believed that God was granting the light of new revelation to the saints, but they, like Hobbes, held that the church could do with further enlightenment. Some also attacked the universities, though their motives were not quite the same as his.

In the opinion of the millenarian John Canne, the universities were the nurseries not of Christ's but of Aristotle's and the state's ministers, and served no useful spiritual purpose. The most important traditional function of the universities was to train the clergy. But radicals often argued that it was God's spirit and not human learning which qualified people to preach. Many urged that the universities be reformed, and in 1653 it was rumoured that the Barebones Parliament – which included a number of radical puritan members – intended to abolish them altogether. On most points, Hobbes' ideas were diametrically opposed to those of extreme puritans, but his opposition to divine-right tithes, his contempt for clerical pretensions, and his criticisms of the universities echoed the views of the radicals. Arguably it was this, as much as anything, which alienated Hobbes' former academic admirers from him.[50]

On some questions, then, Hobbes' religious views coincided with those of puritan radicals; on more he adopted standard Anglican ideas – for instance in response to Bellarmine. His minimalist interpretation of Christian fundamentals, and his arguments in favour of tolerating harmless though heterodox opinions, are close to attitudes expressed by members of Falkland's Tew Circle. On much else, however, he diverged sharply from them. Henry Hammond – a close associate of Falkland's family – denounced *Leviathan* as 'a farrago of all the maddest divinity that ever was read'.[51] Certainly, there was much unusual theology in Hobbes' work – though it is unclear how seriously we should take it, for he may not have believed that Scripture is in fact God's word. But Hobbes did prove – at least to his own satisfaction – that the bible's political message was the same as that contained in the first two parts of *Leviathan*. This was done by a close analysis of the text of Holy Writ. Perhaps, however, the task was not as difficult as it might seem at first glance. For according to Hobbes' own doctrine of Scriptural exegesis, all passages which were too difficult to understand – for instance because they seemed to contradict what reason had demonstrated – could profitably be 'swallowed whole', or, to put the same thing another way, ignored (Lev 32: 256/195).

Conclusion

Hobbes' political writings are commmonly discussed in isolation from the historical circumstances in which they were produced. Commentators treat his theory as a timeless contribution to an eternal philosophical debate. They often claim that the solutions which he offered to the problems of political philosophy were flawed, and they themselves helpfully suggest improvements. Their focus is less on what Hobbes said or why he said it than on questions of modern political philosophy. Certainly, much that is of importance to present-day concerns may be gleaned from Hobbes. But there are dangers in doing modern political philosophy by way of commenting on the old classic texts (read with little or no reference to their historical contexts). One is that the texts may easily be distorted by anachronistic readings. Another is that the authors of those texts – say Thomas Hobbes or Thomas Aquinas – can too readily come to be treated as authorities, and this may discourage impartial consideration of their words and reasons. Hobbes himself warned against this kind of thing when he wrote that 'words are wise mens counters, they do but reckon by them: but they are the mony of fooles, that value them by the authority of an *Aristotle*, a *Cicero*, or a *Thomas*, or any other Doctor whatsoever, if but a man' (Lev 4: 28–9/15).

Hobbes deliberately avoided citing contemporary authors with whose views he agreed or disagreed. This has encouraged the notions that his theory was largely original, that it bore little relationship to the opinions of his contemporaries, and that we may grasp it simply by reading and re-reading his books. Of course, to appreciate Hobbes' thoughts it is necessary to examine the works in which he expressed them, and we should certainly be wary of attributing to him things that he never said. But it is difficult to grasp his full meaning from the texts alone. Hobbes assumed that his audience would be acquainted with the political and intellectual culture of Europe – and especially England – in the early seventeenth century. Readers who lack such an acquaintance are likely to miss much of the point of what Hobbes had to say. He himself stressed that 'it must be extreme hard to find out the opinions and meanings of those men that are gone from us long ago, and have left us no other signification thereof but their books; which cannot possibly be understood without history enough to discover' their intentions and the 'diversity of contexture' in which they wrote (EL1:13:8). The purpose of this book has been to locate Hobbes' ideas within the context of the intellectual and political climate of his age.

It is sometimes said that the theories expressed in *Leviathan* were formulated in response to wholly new political questions which were raised by the Civil War in England – and that the book put forward a political philosophy which was equally new. Kevin Sharpe, for example, tells us that it was only at the time of the war that there 'arose the fundamental questions concerning why men should subject themselves to government, what rights they had and could or should surrender, what limits there were to authority'. Hobbes, he proceeds, 'was forced to answer questions that would have been, quite literally, unnatural to his predecessors', and consequently he 'wrote the first work of political philosophy in England [sic]' (Sharpe 1989, 67–8; *Leviathan* was not, of course, written in England). It is true that *Leviathan* adds to the ideas set out in *De Cive* and *The Elements of Law*, especially on church-state relations and religion; but on most fundamental points the three books say much the same thing, and the *Elements* is patently a work of political philosophy. Of course, it was written before the Civil War. The basics of Hobbes' political creed had been formulated by the Spring of 1640.

During the 1630s Hobbes associated with the circle of Mersenne and fell heavily under the influence of the scientific thinking of Galileo and others. He came to hold that the universe consisted of matter in motion, and that our perceptions and emotions were to be explained in terms of mechanical causation. He rejected the scholastic ideas that people (and other beings) have essential natures which they strive to fulfil. Scholastics argued that God has built into human nature certain purposes (such as self-preservation, promoting the public good, and worshipping God) which men can perceive by using reason. Hobbes denied the existence of any such transcendent purposes, claiming that reason does not tell us what we ought to do, but merely permits us to calculate how we can achieve what we desire. He argued that people's desires vary in accordance with their physical constitutions and experiences, but that all ordinarily fear death and strive to preserve themselves (for self-preservation is a condition of the fulfilment of whatever other desires they happen to have). Though people may disagree about what things they call good (for 'good' is simply a word they use to indicate what they desire), they can all agree that self-preservation is good. It is therefore possible to deduce a series of objective rules governing human action from the principle of self-preservation. Other theorists also gave weight to self-preservation in their moral and political theories, but they held that it was just one of several fundamental moral rules.

Throughout his working life Hobbes was closely associated with the aristocratic Cavendish family. In 1640 he dedicated *The Elements of Law* to the Cavendish Earl of Newcastle, who at that time was governor to the Prince of Wales and who was soon to spend much of his immense fortune in the service of Charles I during the Civil War. In 1627 Hobbes helped to collect Charles' contentious Forced Loan which was widely regarded as an illegal levy since it infringed the principle that the monarch could not take the property of his subjects without their consent. In 1629 Hobbes published his translation of Thucydides in order (as he later remarked) to set before his countrymen the follies of democratic government. During the 1630s Charles called no parliament and once more raised money without obtaining the subject's consent. When parliament met again in the Spring of 1640, the House of Commons objected vociferously to many recent royal policies, and especially to taxation without consent. Such objections were not to the king's liking, and he soon dissolved the parliament. A few days later Hobbes penned the letter dedicating *The Elements of Law* to Newcastle.

The letter records that Hobbes had already communicated the principles set out in the *Elements* to Newcastle 'in private discourse'; it was at the Earl's command that he put those principles 'into method' by writing the book. The letter also shows that Hobbes viewed his work as something more than a dry academic treatise, for he expressed the hope that the book would 'insinuate itself with those whom the matter it containeth most nearly concerneth' – in other words, the king or his leading advisers (EL xv–xvi). The work circulated quite widely in manuscript and was clearly intended to influence influential opinion. Its message was of the most immediate relevance to contemporary politics. Hobbes argued that sovereignty was indivisible, that the sovereign was not bound by the law of the land, and that subjects held no rights of property against him. The teaching of alternative opinions, he suggested, should be rigorously suppressed. Many members of parliament believed that sovereignty was divisible, that the king *was* bound by the law of the land, and that monarchs could never take the property of subjects without their consent.

When Charles recalled parliament in November 1640 the Commons soon began proceedings against people who had expressed views similar to those of Hobbes, and he then fled to France. There he re-wrote his political theory in Latin as *De Cive*, which he completed in the autumn of 1641. Some months later, civil war broke out in England. Hobbes' political theory was not formulated in response to the war, but on the war's eve. In France he associated with exiled English royalists of the highest rank including the future King Charles II himself. Hobbes' theories closely resembled those of other royalists on a great many points. But during the course of the Civil War many of the propagandists who wrote in favour of the king's cause tried to emphasise Charles' moderation. In particular, they stressed his majesty's respect for the subject's rights of property. Before the war some supporters of royal policy had spoken in rather different tones, and had provided trenchant theoretical justifications of the king's extra-parliamentary levies. Roger Maynwaring was one of these. Hobbes – who identified his doctrine with Maynwaring's – was another.

By deducing duties from the single principle of self-preservation Hobbes was able to lay the foundations for an unusually thoroughgoing absolutism. Some thinkers – for instance Bodin and Filmer – freed kings from subjection to the laws of their realms but argued that they were bound by an extensive set of moral rules (or natural laws), and that their subjects should disobey royal edicts which contravened these rules. Hobbes did indeed argue that sovereigns are bound by natural laws. But he greatly reduced the extent of these laws – and thereby increased the powers of sovereigns.

In Hobbes' theory, the laws of nature were rules calculated to promote peace, and so preserve the individual. One such rule was that people should keep their covenants. Hobbes' talk about covenants is a feature of his system which commentators have found especially puzzling. Some have suggested that the theory can be improved by drastically re-writing what he had to say about covenants, or by omitting all reference to them. He brought in covenants, so the argument runs, because he was under the influence of earlier contract theorists; but his system would work much better without them. This approach misses the crucial point that Hobbes' handling of covenants allowed him to reach the vital practical conclusions that subjects can hold no rights of property against their sovereign, and that sovereigns can never commit injustice towards their subjects. Hobbes argued that covenants are highly likely to become invalid in the state of nature, for there is no coercive power there to ensure that anyone will keep his word, and it would be foolish for me to perform my part of a bargain if I have no reason to think you will carry out yours. So people cannot institute property or other rights in the state of nature. Such rights therefore begin only with the erection of a sovereign and are subject to his will. Since what constitutes a person's property is defined by the sovereign's will, no one can hold property against the sovereign.

In his discussion of covenanting Hobbes drew on ideas current amongst casuists and lawyers – including the concept of just fear and the notion that pacts made under coercion bind. In his treatment of the nature of sovereignty he likewise made use of traditional arguments, and there is a good deal of truth in F. W. Maitland's suggestion that 'Hobbes's political feat consisted in giving a new twist to some well worn theories of the juristic order' (Maitland 1965, 304). His system did not mark a total break from orthodox natural law thinking, nor from conventional royalism. True, in *Leviathan* he argued that the English could now covenant to obey the conquering regime of the Rump, and some royalists saw this as a betrayal of the Stuart cause. Hobbes' defence of *de facto* powers – conquerors – has sometimes been seen an instance of a new and hard-headed attitude to politics that first arose amongst apologists for the Rump in the controversy over the Engagement of 1650. But Hobbes' position was based on principles which he had expressed long before and which had been widespread in the 1640s and earlier. Arguably, it was other royalists rather than Hobbes who innovated in the years after 1649 by insisting on indefeasible hereditary right.

What Hobbes had to say on conquerors was only one of the things that offended more conventional royalists about *Leviathan*. Royalists commonly favoured episcopacy and granted the clergy certain spiritual powers. In *Leviathan* Hobbes stripped clerics of all authority (except as delegates of the sovereign), argued against episcopacy, and put forward some startling interpretations of Scripture. The last two parts of the book have long been the least studied, though this has not prevented the emergence of some radically opposed interpretations of Hobbes' religious ideas. A number of scholars (including Strauss and Polin) have argued that Hobbes was an atheist, though for reasons of prudence he concealed the fact. Others claim that he believed in God, and some that he was a Christian whose ideas were close to those of Falkland and his associates in the Tew Circle.

Certainly, Hobbes worshipped as an Anglican and denied charges of atheism. There is no good evidence that he *disbelieved* in God, and his discussion of church-state relations is very similar to Anglican writings on many points. But most Anglicans (and most Christians) held that people have incorporeal souls as well as bodies, and that just as body and soul are distinct so too are civil and spiritual powers or functions. Hobbes was a materialist who rejected the distinction between body and soul – and the corresponding distinction between civil and spiritual. In *Leviathan* he adopted traditional Anglican arguments against Catholic claims that the pope possesses power to coerce temporal sovereigns, but he extended the arguments to demolish the theories of Presbyterians – and of the Anglicans themselves. In 1651, when he published *Leviathan*, the times were propitious for printing attacks on clerical power, since censorship was lax and the clergy weak. *De Cive* included passages reminscent of orthodox Anglican views on episcopacy and clerical power; but these sections do not cohere with the rest of his argument in that book. Hobbes had already formulated the main lines of *Leviathan*'s theory on the clergy's power before the Civil War. Historical context explains why he did not spell his ideas out in print until 1651. By then Anglicans and Presbyterians had been defeated and a Hobbesian solution to England's ecclesiastical problems seemed possible. When the Anglicans returned to power after 1660 they exacted revenge on Hobbes by preventing him from publishing his opinions and by initiating proceedings against him for heresy.

In some ways, Hobbes' interpretation of Scripture was close to ideas which circulated in the Tew Circle, and which became increasingly influential amongst Anglicans after 1660. Hobbes, like Falkland, Chillingworth and Taylor, argued that the bible says that only a small number of beliefs are absolutely necessary to salvation. Like them, he advocated tolerance of harmless though erroneous religious opinions. Taylor published a famous plea for toleration in 1647, at a time when a number of puritan writers were likewise arguing against intolerant Presbyterianism. But Taylor's theory differed from puritan views in two crucial respects, for he gave the magistrate a duty to enforce uniformity of worship, and he imposed upon subjects a duty to obey the magistrate in all indifferent matters. On both points Hobbes adopted the same position as Taylor.

Hobbes' religious ideas were close to those of the Tew Circle on some issues, yet he attracted charges of heresy, while their views acquired an increasing following among orthodox Anglicans. Of course, he adopted some odd theological doctrines (for instance on the Trinity and hell) but he did not insist on his opinions in these areas – retracting what he had said about the Trinity, and revising his account of hell. Far more significant were his materialism – which was commonly seen as necessarily atheistic – and his sceptical attitude towards miracles, prophecy and revelation itself. Notions similiar to those of Hobbes certainly circulated amongst earlier so-called libertines, though in England they were rarely expressed publicly until the collapse of clerical power in the 1640s.

Hobbes' central doctrines on the origins and nature of political society were formulated by 1640 and were intended to refute the ideas that the King of England was subject to the law of the land and especially that he could never tax without consent. Argued with exceptional force and presented as a series of deductions from first principles, the case was constructed to elicit conclusions of immediate practical relevance to current debate, and to disprove opposing theories. The same goes for what he had to say on religion and church-state relations. Hobbes' arguments are steeped in references to the concepts and claims of his contemporaries. No account of his theory which ignores this can be adequate. And thus much concerning Thomas Hobbes and his historical context.

Notes

CHAPTER 1: HOBBES AND HIS CONTEXT

1. A useful collection of writings by Skinner and of *critiques* of his views is Tully 1988. The debate is briefly surveyed in Wootton 1986, 10–14.
2. A good recent account of Hobbes' psychological theories is in Hampton 1986, 17–24. Hobbes on rhetoric and its political implications are discussed in Johnston 1986; a highly important discussion which locates Hobbes' political thinking in the context of classical and Renaissance rhetorical theory is Skinner 1991. Treatments of Hobbes' scientific ideas include Brandt 1928; Pacchi 1965; Shapin and Schaffer 1985. The relationship between Hobbes' scientific ideas and his political theory is discussed in Watkins 1973; Goldsmith 1966; and Sorell 1988. Surveys of Hobbes' thought include Laird 1934; McNeilly 1968; Oakeshott 1975; Peters 1956; Raphael 1977; Sorell 1986; Strauss 1936; Tuck 1989. A good, brief, recent introduction to Hobbes' political ideas is Malcolm 1991, 530–45. Biographical material is to be found in Malcolm 1981; Reik 1977; Robertson 1905; Rogow 1986; Tuck 1989. The most entertaining and in many ways the best life of Hobbes is Aubrey 1:321–403; some passages which were expurgated from that edition are re-instated in Aubrey 1972, 305–20. Though sometimes mistaken on other figures, Aubrey is frequently uncannily accurate on Hobbes: cf. Malcolm 1988, 43.
3. Eachard 1958, 3, 7; cf. Eachard 1673, 130–4, 145, 149–61. Contemporary reactions to Hobbes' ideas are discussed in Bowle 1951; Mintz 1969; Goldie 1991a. Hampton 1986, 189. Gauthier 1988, 148.
4. Gee 1658 frequently cites Hobbes (e.g. 22, 82, 128, 141) along with a great many other authorities. Gee gives no indication that he thought there was any significant difference in approach between Hobbes and the rest.
5. Hobbes to Waller, 8 August 1645, in Wikelund 266. Skinner 1990, 146 and n162.
6. Descartes 1964–76, 4:67.
7. Richard Tuck 1987 argues that Hobbes' emphasis on self-preservation was anticipated by Hugo Grotius (110–13), and suggests that 'the *leitmotivs* of the modern natural-law school' (a school which, in Tuck's view, began with Grotius and included Hobbes, Selden and Pufendorf) were 'a concern with Carneades [an ancient sceptic whose ideas were revived and developed by such French writers as Montaigne and Charron towards the end of the

sixteenth century], and a commitment to refute what he said by using the principle of self-preservation' (113); see also Tuck 1983. The extent to which Grotius' theory was innovatory has long been debated: e.g. D'Entrèves 1970, 53–6; Luscombe 1982, 719; Trentman 1982, 833; see p. 179n50. That Grotius grounded his moral system on the single principle of self-preservation is difficult to show, though it is true that (like, say, Aquinas and Suarez) he gave much weight to this principle: see p. 177n6. Tuck 1987, 109 affirms that in section 5 of the Prolegomena to his *De jure belli ac pacis* (Grotius 1689, iv), Grotius 'did signal quite clearly that his main intention was to answer the sceptic'; but Grotius in fact brushes the sceptical case aside in two brief passages (Prolegomena sections 5, 16–19; Grotius 1689, iv, xi–xii). In section 28 of the Prolegomena (Grotius 1689, xvii), Grotius states that there were 'many and grave reasons' ('causas . . . multas ac graves') why he wrote the book, but he does not list answering the sceptic amongst these. Rather, he tells us that his main intention in writing the work was to provide the first systematic account of the laws of war, in the hope that by so doing he would prevent people from being killed for no good reason (Prolegomena 28–38; Grotius 1689, xvii–xxi). It is still more difficult to show that Hobbes was particularly concerned with answering scepticism. Though he was not reticent in spelling out exactly what doctrines he was writing against (Aristotelianism, classical republicanism, ideas of mixed monarchy, Coke's views of the law), he said very little indeed about sceptics. A sensitive discussion of Hobbes' relationship with contemporary scepticism is Popkin 1982; Popkin argues that Hobbes was 'almost oblivious to his contemporary epistemological skeptics' (145). Missner 1983 claims that in some senses Hobbes himself became increasingly sceptical between *The Elements of Law* and *Leviathan*. Skinner 1991, contains much valuable material on Hobbes and scepticism. He gives evidence for thinking that Hobbes first read Montaigne only after 1642 (37), and persuasively argues that Hobbes' political ideas were concerned less with rebutting epistemological scepticism than with confronting a different kind of scepticism which arose within the field of rhetoric, and which centred on the rhetorical figure of *paradiastole,* 'the precise purpose of which was to show that any given action can always be redescribed in such a way as to suggest that its moral character may be open to some measure of doubt' (3). Skinner traces the use of this technique in writings of Machiavelli, Lipsius, Montaigne and others. He argues that Hobbes was familiar with ancient and recent literature on the subject, and that he agreed with many of the premisses adopted by those who drew relativistic conclusions from this literature, but that his theory was constructed to avoid complete relativism.

It is true that some figures of the Enlightenment made great claims about the importance and novelty of the moral thinking of Grotius, Hobbes, and other writers whose religious outlook was (like their own) tolerant and anti-clerical. But these claims should be treated with some suspicion since it is evident that the *philosophes* would

have found it embarrassing to admit that much of the moral theory of Grotius may also be found in the pages of that impeccably orthodox Catholic, Suarez.
8. Rogow 1986, 29.
9. A good account of this controversy is in C. H. McIlwain's introduction to James I, 1918, lv–lxxx.
10. There is evidence on Hobbes' attitudes towards (and relations with) women in Nicastro 1973, 15 and n123; Rogow 1986, 130–2, 230–1; *Memorable sayings of Mr. Hobbes*: 'Wealth, like women, is to be used, not loved (Platonically)'.
11. Sir Charles is discussed in Jacquot (1949–50), 81–8.
12. Gabrieli 1957 at 227n41. Robert Mason to Hobbes, Cambridge, 10 December 1622, in Tönnies 1936, 83 mentions Hobbes' letter.
13. Gabrieli 1957, 195–250, 237. De Dominis' translation of Bacon is discussed in Malcolm 1984, 47–60. William Cavendish (second earl of Devonshire) owned a copy of De Dominis' *De radiis visus et lucis* (Venice 1611) on optics – ibid. 54.
14. Hobbes to the Countess of Devonshire, London, 6 November 1628, in Tönnies 1903–4, 291–317, at 291–2, indicates that it was then nearly ready to be published.
15. SP 16/79/67. I owe this reference to Dr Richard Cust, whose *The Forced Loan and English Politics, 1626–1628* (Oxford 1987) is a masterly treatment of its subject.
16. Maynwaring 1627, 2:42; Sibthorp 1627, 23.
17. Rushworth 1659–1701, 1:589.
18. Johnson, Keeler and others 1977–83, 5:435–6; Devonshire argued that the law of the land included the king's prerogative and the Civil Law, as well as the common law. Rushworth 1659–1701, 1:605. Charles I 1628, 2–3.
19. Malcolm 1981 is an excellent discussion of this and other aspects of Hobbes' early career.
20. Wood 1813–20, column 476. Matthews 1948, 2.
21. Aglionby to Hobbes 'at Genova' (i.e. Genoa, or possibly Geneva), dated from Leicester Abbey, November 18 or 8 1629, in Tönnies 1936, 84–6. Proceedings connected with the treatise: Gardiner 1883, 7:138–41.
22. Clarendon 1676, 3. An excellent account of the Tew Circle is Hugh Trevor-Roper, 'The Great Tew Circle', Trevor-Roper 1989, 166–230. Doubts about the extent of the link between Hobbes and the Tew Circle are expressed in Zagorin 1985, especially at 596–600. Aspects of the relationship between Hyde and Hobbes are also discussed in Dzelzainis 1989, which provides evidence that Hyde read the *Elements* not long after it was written.
23. Until recently it was widely accepted that Hobbes had worked out much of his later philosophical system by about 1630. One piece of evidence for this is a manuscript treatise usually referred to as 'A Short Tract on first principles', and tentatively dated by the leading Hobbes scholar Ferdinand Tönnies to 'the year 1630' (printed in EL Appendix I, pp.193–210; dated at p. xiii). But more

recently it has plausibly been claimed that the 'Short Tract' is not in fact by Hobbes, and its precise date is uncertain: Tuck 1988 at 16–18.

24. Hobbes to Sir Gervase Clifton, Hardwick, 2 November 1630, in De Beer 1950 at 205.
25. H. M. C. Portland 2:124.
26. Walter Warner to Robert Payne, 17 October 1634, in CM4:380–1. Payne, the Welbeck group, and Sir Thomas Aylesbury's circle are discussed in Feingold 1985. In a letter to Warner of 3 October 1636, Payne transmitted his thanks to Hyde for favours: B.L. Additional MSS 4458, f.27a. The classic account of Hobbes' scientific thinking and its genesis is Brandt 1928. Much modern scholarship on the topic is incorporated in Tuck 1988. Tuck plausibly argues that Hobbes' philosophy was largely developed in the years immediately after 1637 and that it was to a great extent intended as a reply to the ideas of Descartes. The differences between the thinking of Hobbes and Descartes are discussed in Barnouw 1980.
27. Charles Cavendish to Walter Warner, Welbeck, 2/12 May 1636, in CM6:66.
28. Hobbes to Newcastle, Paris, 25 August 1635, in H. M. C. Portland 2:126.
29. H. M. C. Portland 2:128.
30. CM7:435–6, 8:550–2. Bedford 1979, 48. Letter of Robert Payne to an unnamed correspondent, Welbeck, 26 October 1636, Tönnies 1936, 86–7. The relationship between Hobbes' thinking and scepticism is discussed in note 7 above.
31. Letter of Sir Kenelm Digby to Hobbes, 1 October 1636, in Tönnies 1936, 86. Robert Payne to Warner, Welbeck, 3 October 1636, states: 'I hear Mr Hobbes is expected, with his charge, very shortly' (B.L. Additional MSS 4458, f.27a). Hobbes had returned to England by October 16, for there is a letter of that date from him (at Byfleet) to Newcastle: H. M. C. Portland, 2:130.
32. Digby 1644, 153. Letter of Sir Kenelm Digby to Hobbes, London, 11 September 1636, in Tönnies 1936, 89. Descartes 1984–5, 1:120, 129, 2:110–11 (geometrical method); 1:167–8 (colours).
33. Hobbes to Mersenne (for Descartes), 30 March 1641, in CM10:570, asserts that he explained his 'doctrine on the nature and production of light and sound' to Newcastle and Sir Charles Cavendish in 1630. In the dedication to Newcastle of his manuscript treatise 'A minute or first draught of the optiques' (dated 1646) Hobbes affirms that about sixteen years earlier at Welbeck he had told Newcastle that 'light is a fancy of the mind, caused by motion in the brain' (EW7:468). In a letter of 16 October 1636 Hobbes informed Newcastle that 'the motion is only in the medium, and light and colour are but the effects of that motion in the brain' (H. M. C. Portland 2:130). Replying (in 1656) to accusations of plagiarism levelled at him by Seth Ward and John Wilkins (Ward and Wilkins 1654, 53–4; published without authors' names), Hobbes argued that his teachings on this point were substantially different from those of Descartes (which were

incomprehensible) and of Gassendi and Digby (which were a mere re-hash of ancient Epicureanism) (EW7:340–1). On the significance and origins of the doctrine, two contrasting recent discussions are Tuck 1988, and Smith 1990. Tuck argues that the doctrine was developed in response to the scepticism of Montaigne and Charron, while Smith claims that it was evolved largely in order to answer questions raised by late medieval scholastic science and philosophy.

34. Gassendi 1963, 6–7, 30–1, 36–44. Sorbière to Martel, 1 February 1643, quoted in DC 300.
35. A recent introduction to seventeenth-century absolutist thinking is Sommerville 1991. It is sometimes suggested that before 1640 the term 'absolute monarchy' was used to denote an independent kingdom, and not one in which the monarch's powers were unlimited: Daly 1978, 227–50, especially 240–1; Baumgold 1988, 60. However, the term 'absolute' was also commonly used to mean 'unlimited'. For example, in 1628 John Pym told the House of Commons that Roger Maynwaring had erred by attempting 'to infuse into his Majesty that which was most unfit for his royal breast – an absolute power not bounded by law': Johnson, Keeler and others 1977–83, 3:408. Not long afterwards, Sir Robert Filmer used the term in the same way, contrasting 'limited power' with 'absolute jurisdiction', and asserting that although a true king might consult an assembly of his subjects, he 'still reserves the absolute power in himself': *Patriarcha* 3:4, 2:17, in Filmer 1991, 41, 32. Hobbes put the word to similar use in EL2:1:19; cf. DC6:13 Annot; 6:18.
36. *Constitutions and canons*, sig.B4b.
37. For example Copleston 1964, 57.
38. Robertson 1905, 143; cf. Polin 1977, 182–4.
39. Parker 1640, 32, 34.
40. Filmer 1991, 184; the dating of Filmer's *Patriarcha* is discussed in ibid, viii, xxxii–iv.
41. H.M.C. Twelfth Report, part two (Cowper), 251.
42. In 1662 Hobbes stated that 'many gentlemen had copies' of the book *before* the dissolution of this (the Short) parliament; that there was 'much talk of the author'; and that his life would have been in danger if the king had not dissolved the parliament: EW4:414. If Hobbes' memory was accurate, copies of the book were circulating before the date of the dedication. There is no evidence in surviving records of debates in the Short Parliament that Hobbes' views were discussed there; but similar opinions of Maynwaring and Beale were criticised in the House of Commons: Cope and Coates 1977, 114; Maltby 1988, 9–10, 112–15.
43. Hobbes to Scudamore, Paris, 12 April 1641 (new style), in Zagorin 1978 at 159–60. This letter largely confirms the much later accounts in EW4:414 and Aubrey 1:334.
44. Hobbes to the Earl of Devonshire, Paris, 23 July/ 2 August 1641, in Tönnies 1904, 302. A facsimile of this letter is included in EW1, immediately before p. 1.

45. The mutual hostility between Hobbes and Descartes is a running theme in the correspondence of John Pell with Sir Charles Cavendish: Hobbes to Cavendish, 8 February 1641 (CM10:501–4); Pell to Cavendish, Amsterdam, 7/17 September [1644] (CM13:221); Pell to Cavendish, Amsterdam, 2/12 March [1646] (B.L. Addit. MSS. 4280, f.117b); Cavendish to Pell, Hamburg, 19/29 September 1644 (B.L. Addit. MSS. 4278, f.182a; CM13:228); Cavendish to Pell, Hamburg, October 10/20 1644 (CM13:249); Cavendish to Pell, Hamburg, December 10/20 1644 (CM13:277). In 1644 Descartes visited Paris but Hobbes refused to meet him: Cavendish to Pell, Paris, 1/11 May 1645 (CM13:443). On 6/16 February 1646, Cavendish reported to Pell that 'Mr Hobbes confesses Descartes to be a good geometrician, & says if he had employed his time wholly in it, he thinks he would have been inferior to none', but that as things stand Roberval, Cavalieri, Torricelli and Fermat are better geometers (CM14:65; cf. Aubrey 1:367). Descartes held that Hobbes was better at moral philosophy than at metaphysics, but not very good at either: letter of early 1643 in Descartes 1964–76, 4:67. The two men met through the efforts of Newcastle and Cavendish (who had a high opinion of both) in 1648. They agreed on some points but strongly disagreed on others: Cavendish to Pell, 2 August 1648 (CM16:463). Richard Tuck suggests that by 1646 Hobbes' attitude towards Descartes had become much more friendly (Tuck 1989, 25; Tuck 1988, 25–6); but the evidence for this is Hobbes' manuscript treatise 'A Minute or First Draught of the Optiques', which was specially written for Newcastle and Cavendish. Since these men thought highly of Descartes and since Hobbes stood to benefit from their favour it would have been imprudent for him to express himself too vigorously on the matter of the Frenchman. In May 1646 Hobbes wrote to Sorbière warning him that on no account must he let Descartes discover that a second edition of *De Cive* was being projected, or indeed that there were plans to publish anything by Hobbes, for Descartes would be certain to oppose them (DC appendix B, p. 301); the implication is that Hobbes and Descartes were on terms of bitter hostility in 1646.
46. Pell to Cavendish, 9/19 May 1645 (CM13:448).
47. Cavendish to Pell, 7 December 1646, in CM14:663. Henry Hammond to Gilbert Sheldon, 25 November, B. L. Harleian MSS 6942, f.70a ('oddest fellow').
48. Hobbes to Waller, Rouen, 8 August 1645, in Wikelund 1969, 265–6.
49. Aubrey 1972, 134.
50. Malcolm 1988 at 51 and n30, gives details of Hobbes' contributions to Mersenne's *Cogitata Physico-Mathematica* and *Universae Geometriae Synopsis*, both of which were published at Paris in 1644.
51. Cavendish to Pell, 1/11 November 1645 (CM13:523). The work is 'A Minute or First Draught of the Optiques', B.L. Harleian MSS. 3360.
52. Skinner 1966 at 162.
53. Hobbes' initial response to Bramhall – in the form of a letter to Newcastle – was dated from Rouen, August 20 (EW4:278); in the

course of the controversy Hobbes said that the year of this letter was 1646 (EW5:25); but Bramhall asserted that he first read Hobbes' letter on 20 April 1646, and apologised for the delay in doing so, excusing it on the grounds of 'my journey, and afterwards some other trifles which we call business' (EW5:29–30). Hobbes' only known trip to Rouen occurred in the summer of 1645 (on 27 July 1645 old style Cavendish reported to Pell that 'Mr Hobbes is gone to Rouen' – B.L. Addit. MSS. 4278, f.210a; on 8 August 1645 Hobbes wrote to Waller from Rouen saying that he intended to return to Paris within a fortnight – Wikelund 1969, 266). It may well be, therefore, that Hobbes wrote the letter in 1645, not 1646.

54. A brief description of the controversy with Bramhall is in Robertson 1905, 163–7. The debate is discussed in Laird 1934, 189–96; Leibniz 1985; Van den Enden 1979.
55. Sorbière to Hobbes, The Hague, 21 May 1646, in DC appendix B, 301; Sorbière to Thomas Bartholinus, Leyden, 1 February 1647, in ibid, 308.
56. Skinner 1966 at 161–2.
57. DC 85–6.
58. Sorbière to H. Born, Leyden, 27 March 1647, in DC appendix B, 312.
59. For example Cavendish to Pell, October 1646 (CM14:534); 2 August 1648 (CM16:462).
60. Malcolm 1988, 52–3.
61. Cavendish to Pell, 27 December 1644 (CM13:287–8).
62. Hobbes to Sorbière, Paris, 1 June 1646, in DC appendix B, 302 (cf. Tönnies 1889–90, 69; 'intra annum vertentem' means 'within a year'). Hobbes to Sorbière, 16 May 1646, in ibid, 301 (cf. Tönnies, 1889–90, 68) also refers to his plan to go to Montauban with de Martel. Further references to the plan are in Cavendish to Pell, Paris 9/19 July 1646 (B.L. Addit. MSS. 4278, f.259a) and Cavendish to Pell, Paris, 7 December 1646 (CM14:663). In his Latin prose life, Hobbes records that shortly before he began to teach mathematics to the Prince of Wales he was invited to go to Languedoc with a certain nobleman of Languedoc (LW1: xv). It is often stated that the nobleman in question was François de Bonneau, Seigneur du Verdus (e.g. Robertson 1905, 62; Laird 1934, 13; Rogow 1986, 144; Warrender, introduction to DC 9; Skinner 1966, 156). In 1660 Hobbes dedicated his *Examinatio et emendatio mathematicae hodiernae* to du Verdus, addressing him as a noble from Aquitaine ('nobili Aquitano' – LW4: vii). Du Verdus came from Bordeaux (Robertson 1905, 236; Rogow 1986, 144–5), which is not in Languedoc but in Aquitaine; de Martel came from Montauban, which is in Languedoc.
63. Cavendish to Pell, 12 October 1646 (CM14:534).
64. Hobbes to Sorbière, 27 November 1647, in DC appendix B, 314–15; Tönnies 1889–90, 206–7.
65. Cavendish to Pell, 2 August 1648 (CM16:462).
66. Cavendish to Pell, 1 March 1649/50, in B.L. Addit. MSS. 4278, f.295a.

67. Robert Payne to Gilbert Sheldon, 13 May 1650, in 'Illustrations of the state of the church', 172; 4 February 1649/50 (167); 19 August 1650 (172–3); 7 March 1649/50 (170).
68. Josselin 1976, 251, 303. Lawson 1657, 156.
69. DC, Editor's Introduction, 16n1.
70. The date of Hobbes' return to England is usually given as late in 1651: e.g. Rogow 1986, 151: 'Hobbes disembarked at Dover in November or December 1651'. Clarendon 1676, 8, said that Hobbes fled from Paris a 'few daies before I came thither'. Hyde (Clarendon) arrived in Paris on 25 December 1651: Ollard 1987, 148. The correspondence between Nicholas and Hyde in January 1652 treats Hobbes' exclusion from court as a recent event, and Nicholas first reported news of Hobbes' reception at London on 12/22 February 1652: Nicholas to Hyde, calendared in Ogle, Bliss and Macray 1869–76, 2:122; cf. Nicholas to Lord Hatton, in Warner 1886, 286–7.
71. Warner 1886, 284–6. Hyde quoted in Robertson 1905, 73n1. Hobbes' exclusion from court was still being reported as news on 3 February 1652: H.M.C. seventh report, 458.
72. Ascham 1649, 121. Skinner 1972, 79–98, at 94–5 emphasises the similarities between Hobbes and Ascham while State 1985, 29–32 discusses some of the differences. Skinner somewhat revises his earlier views in Skinner 1990, 145–6.
73. E[utactus] P[hilodemius] 1650, 16. This pamphlet is sometimes attributed to Ascham.
74. See chapter 3, section 2. Clarendon 1676, 8, suggested that Hobbes wrote *Leviathan* because he had 'a mind to go home', and Nicholas in February 1652 reported that Hobbes was 'caressed' at London because of his traitorous opinions: Ogle, Bliss and Macray 1869–76, 2:122. It is true that in his *Six lessons to the professors* of 1656 Hobbes boasted that *Leviathan* had persuaded 'a thousand gentlemen' to obey the established government (EW7:336). But after the Restoration he also observed – with little exaggeration – that 'there is scarce a page' in the book which does not upbraid those who rebelled against Charles I (EW4:414), and he claimed that his remarks on the lawfulness of consenting to the rule of a conqueror were intended only to vindicate royalists who acknowledged the new government as the price of recovering their estates (EW4:424). Hobbes was patently aware of the practical implications of what he said on conquerors in *Leviathan*. It is likely that he hoped his remarks would please England's new rulers, but it is unlikely that he expected them to endorse many of his other doctrines. In 1649 he had contemplated visiting England, but refrained from doing so because political circumstances there were still unstable, and not (apparently) because he feared that the parliamentarians would persecute him: Payne to Sheldon, 25 April 1649, in 'Illustrations', 165. Writing to Sorbière on 22 March 1647, Hobbes protested at the inclusion in the second edition of *De Cive* of a portrait of him with an inscription describing him as tutor to the Prince of Wales, since (amongst other

things) the idea that he was closely associated with the Prince might make difficult his return to England at a later date; moreover, the inscription might harm the Prince by suggesting that he endorsed Hobbes' political principles – which were, Hobbes acknowledged, 'abhorrent to the views of almost everyone': DC 311. Hobbes in 1651 (unlike 1640) does not seem to have been particularly fearful that his political principles would endanger him in England – and in the late 1640s and early 1650s it was, indeed, relatively easy to publish with impunity books expressing ideas very different from those of England's rulers (the royalist publisher Royston issued many such works, including titles by Filmer).

75. Burton 1828, 1:349.
76. Tuck 1989, 31. Malcolm 1988, 58. Aubrey relates that when Selden was dying in 1654, a clergyman wanted to attend him, but Hobbes happened to be present and said '"What, will you that have wrote like a man, now die like a woman?" So the minister was not let in': Aubrey 2:221. The story is denied in Baxter 1696, part 3, 48, and also by Brian Duppa: 'That which you have heard of Mr Selden, of his receaving the sacrament humbly and devoutly upon his knees and from the hands of a regular authorised minister of this poor Church of England, is very true, and possibly the Leviathan himself, when the hook shall be put into his nostrills may do as much': Isham 1951, 107; cf. 102–3.
77. Henry Stubbe to Hobbes, 11 April 1657, in Nicastro 1973, 28. Firth 1895, 107.
78. Duppa to Isham, [April 1651], in Isham 1951, 34–5; 15 July [1651] (41). Barlow to Hobbes, 23 December 1656, in Nicastro 1973, 19–20.
79. Malcolm 1988, especially 57–65. A different account is in Skinner 1969.
80. A good recent discussion of parliamentary proceedings connected with the heterodox ideas put forward in *Leviathan* is Tuck 1990, 157–8.
81. Hobbes also discussed this issue in a briefer manuscript, first published in 1968: Mintz 1968.
82. A brief account is in Robertson 1905, 167–85.
83. Skinner, 'Hobbes on sovereignty'.

CHAPTER 2: THE LAW OF NATURE AND THE NATURAL CONDITION OF MANKIND

1. Malcolm 1991, 534.
2. In *De Homine* 11:6 (LW2:98) Hobbes admits that the pain caused by disease can be so great that people desire death in order to release them from it; he here argues that the greatest of evils is death 'especially with torture' ('praesertim cum cruciatu'). So rational people will *ordinarily* rather that *always* strive to preserve themselves. A good, brief discussion is in Jean Hampton 1986, 14–16.

Notes to pp. 30–37

3. A succinct discussion of the differences between the three accounts of the laws of nature is in M. M. Goldsmith's introduction to EL, at x–xv.
4. The idea that the Golden Rule provides a summary of natural law dates back to medieval canon law: Laird 1934, 179–80n4.
5. Clarendon 1676, 40; Suarez 1971–81, 5:30 (III, iii, 3).
6. ST 1a2ae, q.94, art.2, resp. Suarez 1971–81, 3:128–30 (II, viii, 4) argues similarly. Grotius held that moral truths are deducible from facts about people's 'rational and social nature' ('natura rationali ac sociali': Grotius 1689, 14; I, i, xii, 1); unlike Hobbes, he made no attempt to mount a rigidly deductive moral system based on self-preservation alone.
7. ST 2a2ae, q.64, art.7, resp: 'plus tenetur homo vitae suae providere quam vitae alienae'; 'vim vi repellere licet cum moderamine inculpatae tutelae'. Suarez 1978, 75 (VI, iv, 5); 'ius tuendae vitae est maximum'. Almain 1606b, 687: 'Lege autem naturali quilibet tenetur se conservari in esse'.
8. Bridge 1642, 2. *A cleere and full vindication*, 9. Cook 1647a, 32. Danaeus 1577, f. 209v; Towerson 1676, 346. Rutherford 1843, 178.
9. Lessius 1612, 92; Suarez 1978, 75–6 (VI, iv, 5–6).
10. Parker 1642, 16–17. Prynne 1642, sig. A3b. Almain 1606b, 690; 1606a, 707.
11. Hudson 1647, sig. Z2b-4a. Ferne 1643b, 90, 94.
12. Hammond 1650, 20; Digges 1643, 5; cf. Digges 1642, 45.
13. Grotius 1647, 45: 'Defensio enim violenta cum adversus parem sit licita, adversus superiorem illicita est'. Grotius mentions a work by Thomas Preston (alias Roger Widdrington) which was written and published in 1616 (23), so his book cannot date from before that year; nor can it date from after 1617, when Grotius sent Lancelot Andrewes a copy: Andrewes 1854, lxxxvii–xciv.
14. Owen 1622, 76–80. Preston 1612, 31: 'At contra Magistratum publicum, qui innocentes vexat, opprimit, privatque iure suo, non eandem se armis defendendi licentiam idem ius naturae, aut gentium permittit, sed recto, ac legitimo ordine servato procedendum est, ne in republica graviora mala sequantur, & per tumultus, ac seditiones communis civium tranquillitas inique perturbetur'.
15. Barclay 1600, 159 (book 3, chapter 8).
16. Locke 1988, 419–20 (II, 232–3); cf. Nedham 1659, 34; Bridge 1643, 28–9; Salmon 1959, 85, 87, 104, 137.
17. Clarendon 1676, 87. Bramhall 1658, 513.
18. A good recent *critique* of Hobbes' doctrine which argues that the royalist criticism was largely justified is Jean Hampton 1986, 197–207. Leibniz was also dissatisfied with Hobbes' position on this point: Leibniz 1972, 61.
19. Tuck 1979, 121–3 argues that in *The Elements of Law* (unlike *De Cive* and *Leviathan*) Hobbes adopted the usual royalist view that people renounce the right of self-defence when they enter the state. It is true that in the *Elements* Hobbes claimed that individuals give up the right to *resist* the sovereign (EL1:9:10; 2:1:7; 2:1:19); but he

continued to maintain this position in *De Cive*, arguing that in entering civil society people bind themselves not to resist ('ad non resistendum') the sovereign's will (DC5:7); there he pointed out that they nevertheless retain the right of self-defence (ibid), and in the *Elements* he also insisted that people cannot lay down the right to defend themselves (EL1:17:2). The doctrine of the *Elements* is that people always have the right of self-defence, but that in entering civil society they give up the right of resisting the sovereign in cases where self-defence is not directly involved. This is compatible with his later views but incompatible with conventional royalism.

20. There is a large modern literature on Hobbes' doctrine of the state of nature; good discussions include Richard Ashcraft 1971 and François Tricaud 1988.
21. Gerson 1606b, 192 ('de statu naturae lapsae'). Skinner 1978, 2:156–7 (Molina). Grotius 1689, 239 (II, v, ix, 2): 'in statu naturali'.
22. Suarez 1971–81, 5:9–10 (III, i, 3–4): 'inter se divisae, vix posset pax inter homines conservari'; 'summa confusio'. Bolton 1635, 10. Symons 1658, 27. Hooker 1977–81, 1:99 (I, x, 4).
23. Taylor 1851, 280 (2:1:1:2, note).
24. Digges 1643, 14.
25. Taylor 1851, 279–80 (2:1:1:2).
26. Selden 1640, 44–5. Pym 1641, sig. 3E2a.
27. Tenison 1670, 133.
28. Sommerville 1986b, 48–9.
29. The detailed variations in the accounts of the state of nature given in the *Elements*, *De Cive* and *Leviathan* are discussed in Tricaud 1988.
30. Filmer 1991, 3 (*Patriarcha* 1:1). A discussion of the thesis of natural equality is in Hampton 1986, 24–27. The idea that people were politically equal in the state of nature is implicit in contractualist thought; for the state of nature is precisely that state in which no one has political authority over anyone else, and in which all people are therefore politically equal; there may indeed be physical and other inequalities in the state of nature, but they make no political difference.
31. Examples of the view that coercive authority dates from the Fall are discussed in Sommerville 1986b, 18–19.
32. Descartes to an unnamed cleric, about 23 February 1643, in Descartes 1964–76, 4:67: 'tres-mauvaises et tres-dangereuses, en ce qu'il suppose tous les hommes méchans, ou qu'il leur donne suiet de l'estre'. Clarendon 1676, 28–9; cf. Eachard 1958, especially 62.
33. Calvin 1975, 1:234–5 (II, ii, 13).
34. Clarendon 1676, 28; cf. Templer 1673, 67. A typical statement on the relationship between self-preservation and other moral principles is in Carpenter 1629, 59: Carpenter insists that we should not commit suicide, but adds that 'the sweetness of life ought not to share so great a moyty in our affections, as to shut out our obedience, when either Religion stands at stake, or our Country craves our assistance, or Iustice challengeth her prerogative'.

35. Tenison 1670, 136–8, 135. Stillingfleet 1661, 32, 15.
36. Selden 1640, 45, 109, 92–4, 118–19, 81, 38. 'natura aut per se justum aut injustum'. Stillingfleet 1661, 32. Selden 1650, 2–3: 'Utrum autem Praefecturae fuerint illis tunc temporis Juridicae, tametsi nulla omnino restarent earundem in sacris literis alibive vestigia, non magis esset dubitandum, quam utrum in societatem vitae Civilem coalescerent tunc ipsi atque animalia, ut genus humanum reliquum'.
37. Taylor 1851, 280, 295 (2:1:1:2 and 34).
38. Donne 1984, 40 (I, i, 7; the edition of 1700, p. 11, reads 'a hundred' for 'abundant'). Taylor 1851, 294, 299–300 (2:1:1:33, 43).
39. Justinian 100 (I, ii, 1).
40. Donne 1984, 43 (I, i, ix). Similar points are made in e.g. Taylor 1851, 287 (2:1:1:15–16); Tenison 1670, 140.
41. Hooker 1977–81, 1:91 (I, viii, 11).
42. Hayward 1603, sig.B1a.
43. Grotius 1689, 14–15 (I, i, xii, 1–2): 'certe probabiliter admodum'; 'potentissima probatio est, si in id quod dicimus omnes consentiant'.
44. Hooker 1977–81, 83–4 (I, viii, 3).
45. Taylor 1851, 290 (2:1:1:22). Gassendi, who also took scepticism very seriously, similarly argued that though honey seems sweet to us we cannot know that it is so in fact, and that science cannot establish what things are like in themselves: Gassendi 1963, 436–7.
46. Selden 1640, 80–1.
47. Grotius 1689, 9–10 (I, i, x, 1): 'Jus naturale est dictatum rectae rationis, indicans actui alicui, ex ejus convenientia aut disconvenientia cum ipsa natura rationali, inesse moralem turpitudinem, aut necessitatem moralem'. Hooker 1977–81, 1:79 (I, viii, 4). Donne 1984, 39 (I, i, 6). Suarez 1971–81, 3:68 (II, v, 10).
48. Coke 1662, sig. b1b-2a. Hobbes' teachings on reason are discussed in Polin 1977, 26–52.
49. Hooker 1977–81, 85, 90 (I, viii, 5 and 9). Suarez 1971–81, 3:131–3 (II, viii, 6–7).
50. Gregory of Rimini, *In librum secundum sententiarum*, dist. 34, quaest. 1, art. 2 (quoted in Suarez 1971–81, 3:80n197: 'Nam si per impossibile ratio divina sive Deus ipse non esset, aut ratio illa esset errans, adhuc, si quis ageret contra rectam rationem angelicam vel humanam aut aliam aliquam si qua esset, peccaret'. Grotius famously said much the same thing: Grotius 1689, viii (prolegomena 11). Discussions of the doctrine include St Leger 1962; Martineau 1935. The fullest account of the background to Hobbes' voluntarism is Malcolm 1982.
51. William of Ockham, *Super quatuor libros sententiarum*, lib. 3, quaest. 19, ad 3 et 4 (quoted in Suarez 1971–81, 3:81n203). Gerson 1606b, 171.
52. Suarez 1971–81, 3:100, 95–6, 105–6, 84–5 (II, vi, sections 17, 13, 23, 5).
53. Suarez 1971–81, 1:23–4 (I, ii, 4).
54. Justinian 97 (I, i); ST 2a2ae, q.58, art.1; Gerson 1606b, 191.

55. For example Laymann 1643, 260 (III, i, iv, 1): 'dicitur commutativa, quia in commutationibus, & potissimum in contractibus occupatur'.
56. Aristotle 1934, 266–7 (V.ii.12–13; 1130b30–1131a9). Gerson 1606a, 144 (fraud). ST 2a2ae q.63 ('acceptio personarum'), art. 2 (the bishop).
57. Quoted in Spelman 1978, 1:258. According to a Jacobean 'Discourse concerning the prerogative of the Crown', 'the declaring of *meum* and *tuum* . . . is the very object of the laws of England, for distributive justice, as it is delivered by Aristotle, cometh not within our laws': Fussner 1957, 206. Fussner's attribution of this work to Camden is doubtful.
58. Laymann 1643, 260 (III, i, iv, 1): 'communis ratio ac proprietas iustitiae particularis magis reperitur in commutativa, quam distributiva . . . Illa respicit ius strictum, sive proprie dictum debitum; haec autem ius ac debitum late acceptum. Strictius enim, & magis proprium ius quisque habet ad petendum rem sibi debitam, v.g. ex contractu emptionis, mutui, alteriusve commutationis'. Grotius 1689, 6–7 (I, i, viii, 1): 'justitia expletrix, quae proprie aut stricte justitiae nomen obtinet'. Macpherson 1964, 63, argues that Hobbes' discussion of commutative and distributive justice 'suggests that he was deliberately rejecting the model of a customary status society' in favour of his own model which was based on the possessive market society then coming into being in England. But there was nothing particularly English about Hobbes' argument, and it is far from clear that his outlook had much to do with the supposed advent of 'possessive market society'. The notion that Hobbes was a distinctively *bourgeois* thinker is difficult to sustain. As Clarendon observed, his claim that individuals hold no rights of property against the sovereign threatened to undermine trade: Clarendon 1676, 99. An excellent analysis of Hobbes' social ideas, which takes issue with Macpherson's views, is Thomas 1965. Strauss 1936, 44–58, 108–28, claimed that in his early accounts of human nature Hobbes emphasised aristocratic values such as courage and honour, but that he later placed greater stress on *bourgeois* virtues; a succinct *critique* of this thesis is in Goldsmith 1966, 244. Ryan 1988, 100–105, convincingly argues (against Macpherson and others) that Hobbes' individualism was not 'capitalist, bourgeois, or privatized' (100). A recent discussion of Hobbes' views on justice (which has little to say about the historical context, however) is Raphael 1988.
59. ST 2a2ae, q.148, art.6, resp.; Ames 1639, ii. 79; Du Boulay 1970, 81.
60. Bramhall 1658, 464, 569–70.
61. Tenison 1670 137. At Lev 12: 81/56 Hobbes does imply that 'Sodomie' is a 'vice', but this apparently only applies where it is also 'against Law'. In EL2:9:3 Hobbes argues that sovereigns have a duty 'to increase the people' and therefore to forbid polygamy, 'marriages within certain degrees of kindred and affinity', and 'such copulations as are against the use of nature', since all of these militate against 'the improvement of mankind'. Presumably the idea is that in order to ensure their own security sovereigns should make laws encouraging the propagation of the species. But

it is far from clear that sovereigns will always or even usually be secured by a continuous rise in the number of their subjects.
62. A good recent introduction to political thinking in the English Civil War is Sanderson 1989. An older treatment (which does, however, discuss the 1650s as well as the 1640s) is Zagorin 1954.
63. Laymann 1643, 370 (III, iv, 1).
64. The oath *ex officio* and the arguments of its critics are discussed in Maguire 1936.
65. Gardiner 1883, 8:332.
66. St German 1974, 228; cf. Finch 1627, 34. Grover 1980 stresses the affinities between Hobbes' ideas on contract and those of St German.
67. Fulbecke 1618, f.6a, f.9b. François Connan, *Commentariorum Iuris Civilis Libri X*, 1:6, 5:1, cited in Grotius 1689, 348 (II, xi, i, 1–2). Grotius 1689, 349–51 (II, xi, i, 4–5). ST 2a2ae, q.88, art.3 ad 1; Sanchez 1654, 11 (book 1, disp. V, dub.3, art 18).
68. Sanchez 1654, 11 (book 1, disp. V, dub. 3, art. 17): 'Dicitur autem promissio simplex, seu nuda, seu pactum nudum, quia nec roboratur ex eo, quod res promissa sit alias debita, nec onus imponitur, nec firmatur iuramento, nec vestitur stipulatione, aut traditione rei'.
69. Grotius 1689, 348 (II, xi, 2).
70. Sarpi 1607b, 180–1. Sanderson 1686a, 78 (IV, xv): 'gravis & justus est metus, & qui cadere potest in constantem virum'; Azorius 1602, 17 (book 1, cap. x): 'cadens in constantem virum'.
71. An excellent recent discussion – which replaces all earlier treatments – of Hobbes' account of liberty is Skinner 1990. Skinner shows that in *Leviathan* Hobbes' 'basic doctrine' on this question is that '[a] free agent is he who, in respect of his powers or abilities, "can do if he will and forbear if he will"' (123–4); Hobbes does indeed speak of laws as binding people from acting, but these bonds are not 'impediments to liberty in the proper signification of the word' (132), though they do restrict '"the Liberty of Subjects"' (132), since '[t]o live as a subject is, by definition, to live in subjection to law' (133–4). Skinner emphasises the differences between the account of liberty presented in *Leviathan* and in Hobbes' earlier works, and suggests that the changes were linked to his polemical aims (140–51), and to his increasing awareness of the dangers presented by 'the classical republican theory of liberty espoused by so many of his fellow-countrymen'. This classical republican theory included the doctrine that liberty is truly secured only in a republic (140). Arguably, however, the differences between Hobbes' various discussions of liberty should not be exaggerated. On Hobbes' attitude towards classical republicanism, Skinner quotes the trenchant passage from *Leviathan* where Hobbes decries the anti-monarchical sentiments of classical authors and concludes that '"there never was anything so deerly bought, as these Western parts have bought the learning of the Greek and Latine tongues"' (140–1; Lev 21: 150/111). These views may perhaps best be seen as an extension rather than a revision of Hobbes' earlier opinions, for in EL2:8:10 he castigates the

doctrine of 'Seneca and others, so greatly esteemed amongst us' that subjects may sometimes kill their sovereigns, concluding that 'this doctrine proceedeth from the Schools of Greece, and from those that writ in the Roman state, in which not only the name of a tyrant, but of a king, was hateful'. Skinner argues that in the *Elements*, Hobbes endorses the position that people who act under coercion are not free agents, citing EL:2:3:2, in which Hobbes contrasts 'the position of a man who "submitteth to an assailant for fear of death" with that of someone who makes a "voluntary offer of subjection"' (149). It is only in *Leviathan*, Skinner claims, that Hobbes 'unequivocally insists that, when a man submits to a conqueror to avoid the present stroke of death, his act of submission is the willing act of a free man' (150). In EL1:15:13, however, Hobbes asserts that 'there appeareth no reason, why that which we do upon fear, should be less firm than that which we do for covetousness. For both the one and the other maketh the action voluntary. And if no covenant should be good, that proceedeth from fear of death, no conditions of peace between enemies, nor any laws could be of force; which are all consented to from that fear'. Hobbes' position is that valid covenants must be entered into voluntarily, and that covenants made through fear of death are of this kind. This appears to be the same doctrine as that presented in *Leviathan*. It also entails that submitting 'to an assailant for fear of death' is a voluntary act (though perhaps a voluntary response rather than a voluntary offer).

72. Cicero 1913, 32–5 (I, x, 32); Ames 1639, iii. 227; Fulbecke 1618, f.5b; Filmer 1991, 43 (*Patriarcha* 3:7).
73. Azorius 1602, 24 (I, xi): 'Quaeritur, An iusiurandum metu extortum, quo alteri aliquid promittitur, valeat; Respondeo, Ex communi sententia Pontificii iuris Doctorum, & Theologorum, valere . . . Is qui iureiurando promittit latroni pecunias, ne occidatur ab eo, debet iusiurandum servare, & promissum exoluere'; 'simplex enim latroni, tyranno, aut usurario facta promissio iure quidem naturali, aut divino valet'.
74. Sanchez 1654, 324 (IV, i, 3); 345 (IV, viii, 4). Laymann 1643, 382 (III, iv, 6, 2); cf. 379 (III, iv, 4, 9); 23 (I, ii, vi, 7). Cf. ST 1a2ae, q.6, art. 6.
75. Grotius 1689, 354 (II, xi, vii, 2): 'Ego omnino illorum accedo sententiae, qui existimant, seposita lege civili, quae obligationem potest tollere aut minuere, eum qui metu promisit aliquid obligari'. Aristotle 1934, 118–19 (III.i.5–6; 1110a8–20). Sanderson 1686a, 78–9 (IV, xv); Taylor 1864, 639–40 (4:1:7:9).
76. The analysis of Hobbes' views on covenants has given rise to some excellent attempts to salvage the theory by revising it, particularly in the light of games theory. Two fine though contrasting examples are Gauthier 1969, and Hampton 1986. Gauthier 1988 takes issue with some of Hampton's contentions.

Amongst the difficulties with the account which Hobbes gives are:

(1) The problem of the fool: Hobbes claims that once a sovereign

has been instituted there are no grounds of 'just fear' that others will fail to abide by their covenants. The same, he says, is true in the state of nature if one side has already performed his part of the bargain. In these two cases, therefore, the law of nature requiring that we keep our covenants becomes operative. But Hobbes notes that 'The Foole hath sayd in his heart, there is no such thing as Justice . . . seriously alleaging, that . . . to make, or not make; keep, or not keep Covenants, was not against Reason, when it conduced to ones benefit' (Lev 15: 101/72). Hobbes' response to the fool is that rational self-interest cannot warrant breach of contract in the two relevant circumstances. He argued that it is irrational to break covenants even if there are prospects that we will benefit by such action, for we cannot bank on this outcome: 'when a man doth a thing which . . . tendeth to his destruction [for example, breaks a contract], howsoever some accident which he could not expect, arriving, may turne it to his benefit; yet such events do not make it reasonably or wisely done' (Lev 15: 102/73). He illustrated his argument with an example taken from Sir Edward Coke. Coke claimed that if the heir to the crown treasonably killed the king, the sentence of treason would automatically become void since the heir would now be king. Hobbes observed that Coke's reasoning constituted an invitation to heirs to kill monarchs, and responded by claiming that rebellion against sovereigns is not rational since success 'cannot reasonably be expected' (Lev 15: 102–3/72–3). Arguably, this response is weak since though rebellion in general may be a risky business it is unclear that this is always so, and it is plausible that heirs may quite frequently reasonably expect to go undetected if they kill kings. In any case, Hobbes said little to demonstrate that there are no circumstances in which we can rationally expect to benefit from breaking covenants in civil society. Discussing the case of a covenant in the state of nature in which one side has already performed, Hobbes argues that it is rational for the other also to perform since failure to do so will signal that you are untrustworthy and people will therefore not permit you to join their confederacies – with the result that your security will be endangered (Lev 15: 102–3/73). A minor objection here is that according to Hobbes' rules, first performance is irrational (Lev 14: 96/68). A more serious objection is that if confederacies of trustworthy people are possible in the state of nature it is difficult to see why they could not combine together to destroy the untrustworthy, and to introduce property and activate the laws of nature without setting up an absolute sovereign. On the other hand, if such confederacies are not possible, then there is no point in signalling my trustworthiness to others, who are likely to use it to my disadvantage. An interesting discussion of Hobbes' ideas on confederacies is in Tarlton 1978, 312–13.

(2) The problem of first performance in the covenant instituting government: Hobbes tells us that in the state of nature it is irrational to perform your part of a covenant first. But people who covenant together to institute a sovereign are still in the state of nature

until at least some start to perform the covenant (by obeying the sovereign), and so empower him to coerce others into obedience. A recent discussion of this difficulty is in Tuck 1989, 68.

(3) Covenants and conquest. Hobbes claims that a covenantual relationship of sovereignty and subjection can arise if a person is conquered and agrees (in return for life, and liberty from bonds) to obey the conqueror (this is discussed below in Chapter 3, section 2). The difficulty is in seeing what grounds the conqueror has for trusting the conquered to keep the covenant. True, if the conqueror already has subjects they may coerce the newly vanquished individual into obedience. But this will not apply in the case of a man embarking on a career of conquest, and taking his first captive. The conqueror must sleep and while doing so his captive may well calculate that it is in his interests to kill (or conquer) the conqueror. Since the conqueror has no good grounds for supposing that the captive will not act in this way, it will be rational for him (before going to sleep) to kill or chain him, thus terminating their relationship of sovereignty and subjection. If it is rational for the conqueror to kill his first captive then he may experience difficulties in acquiring a plurality of subjects by conquest. Hobbes assimilates the family to commonwealths which arise by conquest. Arguably, it is rational for parents in the state of nature to kill (and perhaps eat) their offspring. For as parents age they will come to lack the physical power to coerce the child. When this happens, they will have no grounds to suppose that the child will continue to obey them or will refrain from killing them – since covenants without the sword are not binding. By nurturing the child in infancy they are therefore undermining their own long-term security. Of course, if there are a number of children, one may be coerced into obedience by the rest. But before parents have a number of children they usually have just one, and they cannot be sure that they will later have others. So seemingly the rational course is for parents to kill the first-born child (and presumably also to obtain useful nourishment by eating it, perhaps after appropriate fattening). In EL2:4:3 (cf. DC9:3; Lev 20: 140/103) Hobbes argues that if your children had not given you a (tacit) promise of obedience it would indeed be rational to let them perish rather than to be in danger of being killed or subjected by them when they are grown. But it is difficult to see what grounds the parent has for supposing that the offspring will abide by the promise of obedience once the parent has weakened to the point where (s)he is no longer capable of coercing (and perhaps not even of protecting) the child.

CHAPTER 3: THE ORIGINS OF GOVERNMENT AND THE NATURE OF POLITICAL OBLIGATION

1. Tierney 1982, 26. Le Bret 1632, 3–4, 9–10. Richer 1692, 80–1, 155, 159. Pierre Grégoire and Sir John Hayward argued similarly: Salmon 1991, 234, 249.
2. Grotius' doctrine is discussed in Tuck 1979, 62–3.
3. Discussing the duty to serve as a soldier, Hobbes argued that 'there is allowance to be made for naturall timorousnesse, not onely to women, (of whom no such dangerous duty is expected,) but also to men of feminine courage . . . But he that inrowleth himselfe a Souldier, or taketh imprest mony, taketh away the excuse of a timorous nature; and is obliged, not onely to go to the batell, but also not to run from it, without his Captaines leave': Lev 21: 151–2/112. Later in *Leviathan* he asserted that soldiers are obliged to serve a defeated sovereign provided that he still has an army in the field, though other subjects may covenant to obey the sovereign's enemies: Lev 'Review and Conclusion', 484–5/390. It is unclear that these principles flow very naturally from Hobbes' premises, but it is worth noting that at the time he was completing *Leviathan* Charles II had an army in the field (which was not finally defeated until 3 September 1651), and it is arguable that Hobbes had no wish to be read as encouraging it to desert. Hobbes' theory on the obligations of soldiers is discussed (with little reference to the historical context) in Baumgold 1983 and Baumgold 1988, 87–93.
4. Parker 1642, 2. Bennet 1649, 8. *Vox militaris*, 1647, 3–4.
5. Azorius 1606, 686a. Suarez 1978, 88–9 (VI, iv, 17).
6. An interesting recent discussion of Hobbes' doctrine of authorisation is in Gauthier 1988, 148–52. Gauthier has little to say about the historical context. An event which gave especial impetus to the idea that sovereignty in England is held coordinately by King, Lords and Commons was the publication in 1642 of Charles I's Answer to the Nineteen Propositions; important discussions include Weston and Greenberg 1981, and Weston 1991, 395–404.
7. Tierney 1982, 24.
8. Quoted in Filmer 1991, 74.
9. Parker 1644, 18; Parker 1642, 34, 14. Herle 1642, 18.
10. Collins 1617, 531.
11. Sommerville 1990, 243. Wootton 1986, 45–51.
12. A different interpretation is in Gauthier 1969, 114–15, and Skinner 1990, 149n177. Gauthier argues that in Hobbes' theory the covenant between conqueror and vanquished is 'technically degenerate' and not a true covenant but 'a promise made on certain conditions' (i.e. the vanquished promises to obey the victor provided that he allows him life and liberty, but the victor undertakes no obligations). To substantiate this interpretation, Gauthier refers to the following passage: 'Nor is the Victor obliged by an enemies rendring himselfe, (without promise of life,) to spare him for this his yeelding to discretion; which obliges not the Victor longer, than in his own

discretion hee shall think fit' (Lev 20: 141/104). Gauthier comments that 'to be obliged at one's own discretion is not to be obliged at all' (114), concluding that a sovereign by conquest may kill or enslave the vanquished at his pleasure.

But the quotation has little bearing on the question of the relationship between subject and sovereign, for – as Hobbes goes on to point out almost immediately afterwards – someone who yields himself to the victor not 'on condition of life, but to discretion' is a slave and not a servant or subject, and is under no obligation to the conqueror (Lev 20: 141–2/104: 'he that hath Quarter, hath not his life given, but deferred till farther deliberation; For it is not an yeelding on condition of life, but to discretion. And then onely is his life in security, and his service due, when the Victor hath trusted him with his corporall liberty'). So if the conquered person yields 'to discretion' no obligations arise on either side. But if the victor does promise to trust the vanquished with life and liberty on condition that the latter obey him in all things, then both sides will seemingly acquire obligations. Firstly, the conquered person will have to obey the victor in all things provided that the latter continues to preserve his life and liberty; this is the same obligation that subjects have to sovereigns by institution. Secondly, the sovereign will have the (unenforceable) obligation to preserve the subject's life and liberty provided that the subject obeys him. At first glance this might seem to give the sovereign by conquest fewer powers than the sovereign by institution (who undertakes no obligations towards his subjects), and so to contradict Hobbes' doctrine that sovereigns of all varieties have the same powers. Moreover, if a sovereign by conquest promises to preserve his subject's life and liberty (in return for the latter's absolute obedience), then he seemingly limits his right of nature; but Hobbes asserts that sovereigns retain the right of nature and that it is in virtue of this that they are empowered to punish their subjects (Lev 28: 214/161–2). However, Hobbes defines punishment as *an Evill inflicted by publique Authority, on him that hath done, or omitted that which is Judged by the same Authority to be a Transgression of the Law; to the end that the will of men may thereby the better be disposed to obedience*' (Lev 28: 214/161). Punishment, then, is for disobedience to the sovereign. But if a subject refuses to obey his sovereign by conquest he violates his covenant and the sovereign may punish him: 'For he holdeth his life of his Master, by the covenant of obedience; that is, of owning, and authorising whatsoever the Master shall do. And in case the Master, if he refuse, kill him, or cast him into bonds, or otherwise punish him for his disobedience, he is himselfe the author of the same; and cannot accuse him of injury' (Lev 20: 142/104). On the other hand, the right of nature arguably does not empower a sovereign in either a conquered or an instituted commonwealth to punish obedient subjects, since such action would militate against peace, and thus violate the law of nature – by which the sovereign is bound (Lev 28: 214–15/162). Both kinds of sovereign have the same powers of punishment, and the same obligations laid down by natural law; the victor's promise to spare the vanquished's life in return for total obedience does not make his powers any more limited than those of a sovereign by institution, for in the case of both types of sovereignty the subject has authorised all the sovereign's acts and therefore cannot complain of injury; moreover, it is the sovereign alone who is empowered to determine whether there has been any breach of covenant (Lev 18: 123/89–90).

Gauthier (116) argues that Hobbes 'succeeds in misleading both himself and his readers by accepting the comparison between master and sovereign, servant and subject'. A compelling *critique* of this position is in Hampton 1986, 116–17.

Skinner 1990, 149n177, argues that 'to make the victor a party to the covenant' 'would be contrary to Hobbes's basic contention [Lev 18: 122/89] that "he which is made Soveraigne maketh no Covenant with his Subjects before-hand"'. But Hobbes is here specifically discussing sovereigns by institution, and it is unclear that he intended the precept to apply to conquerors. The finer points of Hobbes' doctrine on conquest and consent are, admittedly, far from pellucid.

13. Discussions of early seventeenth-century views on conquest are in Skinner 1965, 151–78, and Sommerville 1986a (which discusses James I and Saravia at 256). Grotius praises Saravia in Grotius 1647, 50.
14. Ferne 1643a, 11, 31–3, 56; 1643b, 13.
15. Suarez 1856–78, 24:212 (III, ii, 20): 'justa poena'. Suarez 1971–81, 5:41– 2 (III, iv, 4).
16. Jansson 1988, 312, 316. Floyd 1620, 29. Burroughes 1643, 39, 125. Cook 1649, 8.
17. Floyd 1620, 29.
18. Hayward 1603, 22; Grotius 1689, 14 (I, iii, viii, 1); 101 (I, iii, xi, 1); Saravia 1611, 176.
19. Firth and Rait, 1911, 2:325.
20. Skinner 1972a, 92. Skinner's essay remains fundamental to an understanding of the political theory of the Engagement controversy, though some of its findings have more recently been questioned, and Skinner himself has modified his views: Skinner 1990, 145–50. The importance of the idea of a mutual link between protection and obedience in 'the secular case for engagement' is also stressed in Baumgold 1988, 128. The standard account of the pamphlets published in the controversy is Wallace 1964. Other useful discussions of the controversy include Wallace 1968, 43–68; Burgess 1985; and Sampson 1979.
21. Walker 1648, 42. Woodhouse 1938, 66. Chaloner 1646, 6.
22. Ball 1645, 13–14. March 1642, 3: 'sicut subditus regi tenetur, ad obedientiam, ita rex subdito tenetur ad protectionem'.
23. Coke 1658, 587–8, 590, 602 (part 7, Calvin's case).
24. Donne 1610, 240. Sclater 1616, 2. The Convocation book of MDCVI 1844, 51, preface 7–8.
25. Daly 1979, 121n; Preston 1619, 134–5; Barret 1612, 417–18. Ferne

1643a, 32. Dickinson 1619, sig.B2a-b.
26. Sanderson 1674, 110–11. Filmer 1991, 281–6.
27. Hobbes' theory of the family is discussed in Schochet 1975, 225–43; Schochet 1990; Abbott 1981; Chapman 1975; Lund 1988; Pateman 1988, 43–50; Zvesper 1985.
28. Rutherford 1843, 50, 62; cf. Suarez 1971–81, 5:23–4 (III, ii, 3); Bridge 1643, 11–12; Canne 1649, 28; Cook 1647a, 3; Cotton 1645, 4; *Maximes unfolded*, 19.
29. Jackson 1844, 312; Buckeridge 1614, 282.
30. Suarez 1856–78, 24:205 (III, i, 8). Grotius 1689, 234 (II, v, 1): 'Generatione parentibus jus acquiritur in liberos'. Filmer 1991, 192.
31. Towerson 1676, 246.
32. Selden 1616, 49: 'But its cleerly true generally, that where iura conubij were not, there the Roman law makes the issue follow the mother, as the law of nature requires'. Grotius 1689, 262 (II, v, xxix, 1).
33. Commenting on Lev 30: 235/178, William Lucy (Lucy 1673, '99'=69) argued that Hobbes here 'let fall an excellent truth which is clear against the whole Body of his Politicks, which is, that the Fathers of Families, not the Rabble, were the Erectors of Commonwealths'; Schochet 1975, 238–9, makes much the same point, and assimilates Hobbes' teaching to that of Filmer. Certainly, Hobbes claimed that fathers often came to hold power over their families, but he denied the patriarchalist contention that this was always or necessarily the case, and he grounded the father's power not upon the law of nature but upon the consent of the mother and children. It was for these reasons that patriarchalists attacked Hobbes' theory: e.g. Filmer 1991, 191–2; Coke 1662, 26; Templer 1673, 58. Schochet 1975, 239–40, convincingly argues that once Hobbes' doctrine of the family is understood, 'his state of nature no longer appears to have been altogether individualistic'.
34. Warrender 1957, 124, points out that though Hobbes grounds the parent's power upon the child's consent, he elsewhere (Lev 26: 187/140) tells us that children have no power to make covenants. Similarly, Thomas 1965, 188–9, argues that the notion that 'Hobbes regards men as entirely free from relationships based on status' cannot be sustained, since 'there are traces in his writings of patriarchalism which he did not succeed in justifying in wholly contractual terms'; he did indeed claim that the child 'agrees to the power which protects it', but this argument was 'ultimately unsuccessful', since 'as Hobbes himself concedes elsewhere, it is impossible to see how infants can consent to anything'. Hobbes later clarified his doctrine, arguing that: 'The children therefore, when they be grown up to strength enough to do mischief, and to judgment enough to know that other men are kept from doing mischief to them by fear of the sword that protecteth them, in that very act of receiving that protection, and not renouncing it openly, do oblige themselves to obey the laws of their protectors; to which in receiving such protection, they have assented': EW5:180. Children,

in short, acquire consensual obligations only when they grow 'to judgement enough'.
35. Warrender 1957, 8–10; Warrender held that Hobbes himself probably intended to ground obligation 'upon divine reward and punishment', but that he may have wished to ground it 'simply on the will of God' or 'upon a body of natural law having self-evident or intrinsic authority'; he asserted that the last option 'is the solution I would have preferred Hobbes to have taken': Warrender 1965, 90. A.E.Taylor's 'The ethical doctrine of Hobbes' was first published in *Philosophy* for 1938. It is reprinted (with small changes) in Taylor 1965. Brown 1965 contains much useful material on the controversy over the Taylor and Warrender theses. The debate is surveyed in Barry 1972. A recent revival of the dispute is Trainor 1988; and Skinner 1988.
36. Warrender 1957, 278–311.
37. Ibid., 92, 210–13.
38. Ibid., 216, 213; 1965, 97; 1969, 153.
39. Warrender 1957, 28–9, 93–4.
40. Skinner 1972a, especially at 139–41.
41. Filmer 1991, 189.
42. Justinian 391–3 (III, 13), 476–84; Cowell 1607, sig.2Z1b-2a.

CHAPTER 4: HOBBES ON SOVEREIGNTY AND LAW

1. Filmer 1991, 184, viii, xxxii–iv.
2. Locke 1988, 143 (I, 5). Clarendon 1676, 55; Whitehall 1679, 7; cf. ibid, 47, and Whitehall 1680, 70. Sidney 1990, 11: 'The production of Laud, Manwaring, Sybthorpe, Hobbes, Filmer, and Heylin seems to have been reserved as an additional curse to compleat the shame and misery of our age and country'.
3. Bodin and Hobbes are compared in King 1974.
4. Tierney 1982, 30.
5. Bodin is cited in e.g. Cooke 1608, 13–14, 29–30, and frequently elsewhere; Vaughan 1608, sig.R6b; Downing 27, 46–7; Hayward 1606, 13; Morton 1610, 12; Bridges 1587, 787; Ellesmere 1977, 319; Eliot 1879, 2:34 and frequently elsewhere; Spelman 1642, sig. B2b, D3a. Heylin 1658, 145, 233, 239–49, and elsewhere. Bodin's influence in England is discussed in Krautheim 1977.
6. Saravia 1611, 163–4, 173, 122, 119. Other examples of the argument that mixed government is impossible are discussed in Sommerville 1986b, 26. Filmer 1991, 2 (*Patriarcha* 1:1).
7. Adams 1633, 240.
8. Vaughan 1608, sig.R6b; Hayward 1606, 6. The political views of Hayward and other civil lawyers are discussed in Levack 1973, 86–121. James I 1621, 32, 62.
9. Bodin 1606, 153–82 (book 1, chapter 10).
10. Johnson, Keeler and others 1977–83, 3:452, 494.

11. Ibid., 4:182. SP 16/102/14; cf. DC7:17. Kynaston 1629, f.159b, 159a, 172a.
12. Figgis 1914, 5 classes the idea that 'Hereditary right is indefeasible' as one of the four doctrines which together compose the 'theory of the Divine Right of Kings in its completest form'.
13. Bodin 1606, 746-54 (book 6, chapter 5). Bridges 1587, 716, 787; Ridley 1634, 109. Filmer 1991, 7, 44 (*Patriarcha* 1:5 and 3:8). Saravia 1611, 167.
14. Bodin 1606, 715-16 (book 6, chapter 4). Saravia 1611, 174. Filmer 1991, 25-6 (*Patriarcha*, 2:11-12).
15. Filmer 1991, 31 (*Patriarcha* 2:16). Charles I 1640, 19. The political uses to which similar ideas were put in the early Stuart period are discussed in Sommerville 1986b, 134-7; Gunn 1969, 68-82.
16. Bodin 1606, 645 (book 6, chapter 1).
17. For example, Nathaniel Fiennes argued in the House of Commons on 9 February 1641 that if the bishops established 'arbitrary power . . . over their Clergy', 'wee know what an Influence they may have by them upon the people, & that in a short time they may bring them to such blindnesse, and so mould them also to their owne wills, as that they may bring in what Religion they please': *Speeches and passages*, 36.
18. Saravia 1611, 176-8.
19. A recent discussion of Hobbes' attitude towards despotism is Wolin 1990.
20. Barlow 1609, 145, 368. Laird 1934, 72. Filmer 1991, 19 (*Patriarcha* 1:4). Maynwaring 1627, 2:25; Johnson, Keeler and others 1977-83, 5: 621-2.
21. Jackson 1844, 291, 273.
22. Pierre Du Moulin the younger 1640, 37.
23. Filmer 1991, 43-4 (*Patriarcha* 3:7).
24. There is a very large literature on medieval and early-modern theories of property. Useful discussions include Schlatter 1951; Tully 1980; Ryan 1984; Coleman 1988.
25. Gaius 1940, 66.
26. ST 2a2ae, q.66, art.2, resp.: 'necessarium ad humanam vitam'. Ames 1639, iii. 222-3. Cosin 1584, 255.
27. Suarez 1971-81, 4:29-31 (II, xiv, 13-14). Tully 1980, 80, takes Suarez's discussion of property in the '"state of innocence"' to be about 'pre-political society'. However, Suarez, like other authors, means by the 'state of innocence' not pre-political times but times before the Fall.
28. Selden 1652, 19-20, 22. Grotius 1689, 185 (II, ii, ii, 5): 'Simul discimus quomodo res in proprietatem iverint . . . pacto quodam aut expresso, ut per divisionem; aut tacito, ut per occupationem'.
29. Suarez 1971-81, 4:32-5 (II, xiv, 15-17).
30. Eutactus Philodemius 1649, 16. Bodin 1606, 11 (book 1, ch. 2).
31. Woodhouse 1938, 59, 75.
32. Cosin 1584, 255. Cosin did, however, countenance the opinion of the sixteenth-century Catholic theologian Alphonsus de Castro that land was at first held in common.

33. Filmer 1991, 234.
34. Saravia 1611, 178–9, 272 (cf. DC 6:15). Bodin 1606, 665, 222 (book 6, chapter 2; book 2, chapter 5).
35. ST 1a2ae, q.105, art.2, ad tertium.
36. Bodin 1606, 661, 663 (book 6, chapter 2); Saravia 1611, 178.
37. Church 1941, 328–33.
38. Sommerville 1986b, 127–31, 160–3. Suarez 1856–78, 5:479, 493 (V, xiv, 2; xvii, 7; in the English Parliament of 1628 Pym and Mason attacked Maynwaring for misrepresenting Suarez on this point: Johnson, Keeler and others 1977–83, 3:528, 4:108–9. Grotius 1689, 82, 112 (I, iii, 6, 2; I, iii, 17, 1).
39. Foster 1966, 2:157.
40. Pre-Civil War Parliamentarian claims are discussed in Sommerville 1986b, 151–60. The claim that villeinage is unnatural is voiced in Cook 1649, 8–13; Parker 1644, 36–7; *Vox plebis*, 4.
41. Filmer 1991, 19, 209, 225, 236.
42. Cf. e.g. Grotius, *Mare clausum*, chapter 8 (Grotius 1689, 24): 'Dederat natura omnia omnibus'.
43. For example, Fulbecke 1602, f.12b-13a, argued that 'the distinction of demesnes and the property of goods' are by 'the law of nations' but that 'the means whereby they are acquired are prescribed by the civil and common law'; he concluded that since the law of the sovereign could regulate the pre-existing law of nations on the question of acquisition, the sovereign could in emergencies take his subjects' property: 'Princes have for special causes free disposal of their [i.e. their subjects'] lands and goods'. Jeremy Taylor argued that 'all distinction of dominion consists in the sentence and limits of the law' (88), maintaining that the sovereign's will defined legal validity, but not justice, for if the supreme power takes goods 'without just reason, it is *injuste*, but it is *jure factum*' (90): Jeremy Taylor to Dr Richard Bayly, the vigils of Christmas 1648, in Cary 1842, 2:75–99.
44. Justinian 102 (I, 2, 6): 'quod principi placuit, legis habet vigorem'. Bodin 1606, '73'=91 (book 1, chapter 8). James I quoted in Notestein, Relf and Simpson 1935, 2:4. Saravia 1611, 276.
45. Filmer 1991, 57 (*Patriarcha* 3:15). Kynaston 1629, f.153a-b, 159a-b, 164a.
46. Kynaston 1629, f.149b, 152a (on Coke); Hobbes Lev 25: 181–2/135–6 (on parliament). There are some grounds for thinking that Hobbes moderated his hostility towards parliament after the Restoration: Okin 1982.
47. Suarez 1971–81, 5:39–41 (III, iv, 2–3). Grotius 1689, 112 (I, iii, xvii, 1); 8, 228–31, 520–1, 407 (I, i, ix, 1; II, iv, xii, 1; II, xx, xxiv, 1; II, xiv, ix).
48. Suarez 1971–81, 2:53 (I, xi); 5:245–6 (III, xvi, 1); (promulgation); 1:114 (I, vi, 11); 6:263–4 (III, xxxiv, 1) (sanctions); 5:206–8 (III, xiv, 8) (laws must relate to future and not past actions); 1:16–17 (I, i, 7–8) (command and counsel); 1:49–50 (I, iii, 14); 6:53, 57–8 (III, xx, 3 and 8) (law as command of legislator); 5:218 (III, xv, 4): 'voluntas principis sufficit'.

49. Discussions of the thought of Coke and similar lawyers are in Pocock 1957, 30–69; Sommerville 1986b, 86–111. Hale 1924; Hale vindicates the theory of 'artificial reason' at 500–506, and rebuts the idea that the king alone can make law at 506–513; he bases royal power and its limitations upon the subject's consent as witnessed in ancient custom at 505, 507.
50. Similar attacks on the concept of artificial reason by James I and Lord Chancellor Ellesmere are discussed in Sommerville 1986b, 96. The concept itself is discussed in ibid., 89, 93–4.
51. The debate on proclamations is discussed in Heinze 1982, 237–59; and in Sommerville 1986b, 177–8. Like Hobbes, the preacher and scholar William Pemble classed 'edicts and proclamations' as laws, and held that the only unwritten laws were those of nature and nations (though Hobbes equated the laws of nature and nations – Lev 30: 244/185; EL2:10:10): Pemble 1632, 73. Common lawyers regularly regarded their law as unwritten.
52. Bodin 1606, '162'=161 (book 1, chapter 10). Suarez 1973, 161: 'saltem requirit tacitam licentiam principis'. Parsons 1606, 267. Kellison 1621, 291–2; Kellison cites Gabriel Vasquez in support of this position. Filmer 1991, 45 (*Patriarcha* 3:9).
53. Morrill 1980, 27.
54. Cartwright 1573, 22. Cotton 1647, 113; Cotton argued that only extraordinary circumstances could release the magistrate from the obligation to execute murderers.
55. Sommerville 1990, 250–2.
56. Taylor 1647, 199. Suarez 1978, 74 (VI, iv, 4): 'vindicta et poena delictorum ordinantur ad commune bonum reipublicae, et ideo non est commissa nisi ei cui publica potestas gubernandi rempublicam commissa est'. Lessius 1612, 40 (the innocent). Hammond's views are discussed in Tuck 1979, 107–8. A little earlier, the *jus zelotarum* had been discussed by John Selden: Selden 1640, 488–90. For contrasting interpretations of Selden's stance on the *jus zelotarum* see Tuck 1979, 95–6, and Sommerville 1984, 446–7n64.
57. James I 1918, 19–20 (laws); 62, 308 (all subjects equally the king's vassals).
58. The idea that Hobbes' theory (like Locke's) was one of 'early or *classical* liberalism' is expressed by Richard Tuck (amongst many others): Tuck 1989, 73. Tuck states that according to this theory the 'primary responsibility of both citizens and sovereigns is to ensure the physical survival of themselves and their fellow citizens. Once this minimal requirement has been met, policies should not be enforced upon the community' (ibid). But Hobbes says that the office of the sovereign is to secure the safety of the people and adds that 'by Safety here, is not meant a bare Preservation, but also all other Contentments of life, which every man by lawfull Industry, without danger, or hurt to the Common-wealth, shall acquire to himselfe' (Lev 30: 231/175). He argues that the sovereign should intervene in economic life in order to promote 'Navigation, Agriculture, Fishing, and all manner of Manifacture that requires

59. James I 1918, 19.
60. Hampton 1986, 240, argues that there is one passage in *Leviathan* (24: 172/128) in which Hobbes invites 'subjects to judge whether or not the sovereign's policies bring them the security they desire, and to render those policies "void" if the decision goes against them'. She adds that to render the sovereign's policies void must mean at least disobeying them and perhaps even deposing the sovereign (240–1). If this were correct, then plainly the sovereign's will would not be the ultimate criterion of legal validity, at least in the case in question. The relevant passage is (in most copies of the 1651 edition, and in later editions except Richard Tuck's): 'seeing the Soveraign . . . is understood to do nothing but in order to the common Peace and Security, this Distribution of lands, is to be understood as done in order to the same: And consequently, whatsoever Distribution he shall make in prejudice thereof, is contrary to the will of every subject, that committed his Peace, and safety to his discretion, and conscience; and therefore by the will of every one of them, is to be reputed voyd'. However, as Tuck has pointed out (Tuck 1989, 71; cf. Lev p. xxxvi) Hobbes corrected the manuscript of *Leviathan* to read not 'whatsoever Distribution he [i.e. the sovereign] shall make' but 'whatsoever Distribution another shall make'. In other words, if a *subject* distributes lands in a way that militates against the common peace and security, his act is to be deemed void. The ultimate judge of whether such an act is in fact inequitable, and therefore void, is presumably the sovereign. So the passage gives subjects no right to render void their sovereign's commands.

Notes to pp. 103–109

labour' (Lev 30: 239/181). Moreover, he holds that the sovereign 'ought indeed to direct his Civill commands to the Salvation of Souls' (Lev 42: 398/316), and to enforce uniform public worship (Lev 31: 252/192). So Hobbes' sovereign has a duty to provide for far more than the mere preservation of his subjects.

61. Clarendon 1676, passim, and especially 99, 163; cf. Lucy 1673, 43, 217.

CHAPTER 5: HOBBES ON CHURCH AND STATE

1. An illuminating discussion of Bacon's influence and of Hartlib's circle is Webster 1975.
2. Sommerville 1990, 235–6.
3. Ames 1639, iii. 108 argued that 'true right practicall reason, pure and complete in all parts' is to be found only 'in the written Law of God'.
4. Tuck 1989, 85–6; Tuck 1990, 161–3.
5. Wilson 1638, 5. Digby 1638, 49–50. Fulke 1601, 816; Smith 1631, 170–3; Fitzherbert 1615, sig.a4b.
6. Calvin 1975, 1:68–9, 71–2, 74–5, 82–3, 84–5 (I, vii, 1, 4; I, viii, 1, 13; I, ix, 1); cf. Wendel 1965, 158–9. Rogers 1607, 31–2.
7. Hooker 1977–81, 1:232 (III, viii, 15). Chillingworth 1838b, 1:183,

        274 (chapter 1, section 30; chapter 2, section 159). Cheynell 1644, sig. F1b.
8. Chillingworth 1838b, 1:176 (chapter 2, section 24).
9. Here and elsewhere the term clergy is used in its modern sense. Hobbes at Lev 44: 420–2/336–7 rejects the distinction between clergy and laity, and says 'clergy' means those who are to be maintained by tithes, namely Levites. The current distinction between clergy and laity, he says, is a popish innovation based on misconstruing Scripture and intended to further the ambitions of the pope and his followers. Similar arguments occur in e.g. the Presbyterian Gillespie 1645, 36; and the Independent Cotton 1645, 14.
10. Chillingworth 1838b, 2:413 (chapter 6, section 62).
11. Since there was, on Hobbes' view, no sovereign in the country, we might ask why he thought he had to acknowledge as the canonical books of Scripture those writings which had earlier been acknowledged as such by the authority of the church of England: for no such authority now existed. Hobbes claimed that in 1651 he was free to interpret Scripture as he pleased, for there was then no sovereign in England. But he did not claim that he was free to decide which books in fact were Scripture. Why not? A plausible answer is that he thought no books contained God's word, but that he found it convenient to treat as canonical those books which law had earlier authorised, since most of his readers regarded them as divinely inspired. Though Hobbes said he acknowledged the texts which the church of England had endorsed, he sometimes employed the Vulgate rather than the Authorised version of the bible in *Leviathan*: Lucy 1673, 92, 182; Greenleaf 1974, 18n22.
12. Hobbes to the Earl of Devonshire, Paris July 23/ August 2, 1641, in Tönnies 1903–4, 302.
13. Hamilton 1978, 449.
14. Moulin 1614, 311: 'Monarchiam Papalem enatam ex ruinis Imperii Romani'; cf. also '214'=314, 319.
15. Tooker 1611, 148; Buckeridge 1614, 233, 918–23; Warmington 1612, 13; Preston 1611, 282–9; Moulin 1610, 77–8; Field 1847–52, 523 (Athaliah). Tooker 1611, 154–5; Buckeridge 1614, 517–23; Thomson 1611, 37–8 (Solomon). Collins 1617, 531–2; Owen 1622, 12–14; Harris 1614, 218; Burhill 1613, 285–6; Jackson 1844, 237–8 (Ambrose): this attitude to Ambrose's conduct represents a shift from views expressed in Elizabeth's reign, when it was sometimes argued that Ambrose rightly excommunicated the emperor: Collinson 1979. Andrewes 1851a, 444; Burhill 1611, 72; Grotius 1647, 36 (pastors).
16. Bellarmine 1602–3, 1:890 (*De summo pontifice*, book 5, chapter 7). Kellison 1621, 213–17.
17. Bellarmine 1602–3, 1:887–8 (*De summo pontifice*, book 5, chapter 6). Suarez 1856–78, 24:311 (III, xxii, 7): 'temporalia omnia ordinari debent ad spiritualem finem . . . Nam ita subordinantur potestates sicut et fines'.
18. Barclay 1611, 3; Moulin 1610, 61 (indirect and direct power). Moulin 1610, 62 (cooks); cf. Casaubon 1607, 63.

19. Gordon 1610, sig. B3a-b; De Dominis 1617, 43, 47–52; Dove 1610, 8. Buckeridge 1614, 709: 'Regnum meum non est de hoc mundo'; Barclay 1611, 125–6; De Dominis 1620, 493–518; *The Convocation book of MDCVI*, 178–9.
20. Hooker 1977–81, 3:323, 355–6 (VIII, i, 4; VIII, iii, 6).
21. Fitzherbert 1613, 11. *The Convocation book of MDCVI*, 16–18, 26–7.
22. Before the publication of *Leviathan*, Bramhall drew up in manuscript a set of political and theological objections to *De Cive*: Hobbes, *The questions concerning liberty, necessity, and chance*, EW5:24. Bramhall 1658, 556–9, attacks both *De Cive* and *Leviathan* on church-state relations. In 1662 Roger Coke published *Observations on Mr. Hobbs De Cive* (in Coke 1662 at 24–36). Eachard, Tenison, and other critics of Hobbes attacked both *De Cive* and *Leviathan*.
23. Sommerville 1983, 549–51; Downing 1634, 72–3; Heylin 1637, 105. Replying to Bramhall in *The questions concerning liberty, necessity, and chance*, Hobbes took issue with the idea that bishops possessed some powers (including the power of ordination) by divine right, but could exercise them only with the sovereign's permission – 'as if the right to ordain, and the right to exercise ordination, were not the same thing'. To have a right to do something, he claimed, was to be able to do it lawfully, so no one could have a right which he could not exercise (EW5:143).
24. From the 1580s onwards it grew increasingly common for English churchmen to argue that there should ordinarily be bishops in the church, and that bishops derived certain spiritual powers from God alone. These points are discussed with different emphases in e.g. Lake 1988, 88–97; Thompson 1966; Allen 1928, 177–81; M. R. Sommerville 1984.
25. Hobbes' views on ordination resembled those of Independents, who commonly argued that it was election, and not the ceremony of laying on hands, which made a man a minister: e.g. Hooker 1648, sig.b1b, ii.41; Baxter 1696, part 1, 190.
26. In the Latin *Leviathan* (LW3:568) Hobbes says that Elizabeth I did not arrogate to herself the powers of clerics, 'knowing that women are forbidden to speak in churches' ('sciens prohibitum esse foeminis loqui in ecclesiis'). Hobbes does not explain by what law it is that women are forbidden to do this. Presumably it cannot be human law, which would not bind a sovereign queen; nor divine positive law set out in the bible (e.g. 1 Corinthians 14:34), for Christ's mission was not to make laws; nor natural law, for by nature people are equal. Perhaps the injunction that women should not speak in church was just a piece of good advice.
27. Lake 1988, 222–3, 246–7.
28. Selden 1618, 13 (Levites); 10–24 especially 21 (tithes and the civil power); 35–6, 40 (no tithes in early church). Ammianus Marcellinus 1940–52, 3:20 (XXVII, 3, 14).
29. Hobbes to the Earl of Devonshire, Paris, July 23/August 2 1641, Tönnies 1903–4, 307.
30. Coke 1662, 29.

31. Burn 1788, 2:218.
32. Burn 1788, 2:222–34 gives details of the law. Its enforcement is discussed in e.g. Marchant 1969, 220–2; O'Day and Heal 1976, 21–2, 232, 245, 252, 257; Thomas 1973, 312–13.
33. Calvin 1975, 2:441–5 (IV, xi, 3–6).
34. A good, brief account of Erastus and his influence in England is in Figgis 1914, 293–342. Mitchell and Struthers 1874, 197; Coleman 1646, 16 (Coleman). Prynne 1645, sig. A4a, 48 (Prynne). Selden 1650, x; Watson 1651, 61 (Selden).
35. Saravia to Bancroft, 23 August 1590, B.L. Addit. MSS, f.167a. Saravia 1592, 2. Sanderson 1661, 69. Stillingfleet 1661, appendix ('A discourse concerning the power of excommunication in a Christian church'; this appendix is dated 1662), 22.
36. Buckeridge 1614, '577'=576: 'Excommunicationem . . . poenam esse salutarem, norunt omnes, poenam esse coactivam, qua quisquam invitus cogatur, ignorant omnes; quia super voluntatem, non super noluntatem, operatur'. Taylor 1864, 272 (3:4:1:13–14).
37. Rogers 1607, 190. ST 3a suppl, q.23, art.1, resp. Diana 1660, 314 ('Excommunicatio', sect. XXVII).
38. Andrewes 1851b, 58–9; Collins 1617, 531–2; Owen 1622, 12–14; Harris 1614, 218; Burhill 1613, 285–6; Jackson 1844, 237–8. Erastus maintained the same position: Erastus 1589, 300–1.
39. Thomson 1611, 58–60.
40. For example, King 1607, 23; Owen 1622, 9.
41. Erastus 1844, 73 (thesis 39); Erastus 1589, 257 (modern invention); the same claim occurs in e.g. Vedel 1647, 81; Harrington 1977, 217. Erastus 1589, 163–4, 170 (only secular penalties in bible); John Lightfoot made the same claim in the Westminster Assembly: Mitchell and Struthers 1874, 439–40. Erastus 1589, 157–8, 176–9; the ideas that the church in Matthew 18 was the sanhedrin, and that this was a civil court are expressed in e.g. Doughty 1651, 92; Harrington 1977, 184, 371; Selden, speech of 15 September 1645, in Rushworth 1659–1701, 6:203; Selden 1927, 45, 100; Selden 1650, 16–17.
42. Erastus 1589, 191. The same claim is made in Lewis Du Moulin 1650, 86–7. In a letter to Hobbes of 7 October 1656 Henry Stubbe refers to Du Moulin as 'our Du Moulin': Nicastro 1973, 8. This Du Moulin was the son of Pierre Du Moulin the elder, and brother of Pierre the younger.
43. Erastus 1589, 295. The same point is made in e.g. Prynne 1645, 8; Owen 1622, 9; Vedel 1647, 79.
44. Erastus 1589, 258; cf. Coleman 1645, 25: 'I could never yet see, how two coordinate governments exempt from superiority and inferiority can be in one state, and in Scripture no such thing is found, that I know of'.
45. Erastus 1589, 163–4.
46. A number of Anglicans took Erastus's line that Matthew 18:17 is irrelevant to excommunication: e.g. Bilson 1593, 37; Sutcliffe 1591, 54–57; Bridges 1587, 961, 1014–19. But these writers dissented from Erastus in arguing that there was an ecclesiastical censure

of excommunication which was to be exercised by ecclesiastics, and in particular by bishops: Bilson 1593, 104; Sutcliffe 1591, 71, 88; Bridges 1587, 1019, 1080. Other leading Anglicans grounded the power of excommunication on Matthew 18:17: e.g. Hooker 1977–81, 3:14–15 (VI, iv, 1); Rogers 1607, 190.

Hobbes says that Christ forbade his followers to eat with excommunicates; but he also informs us that Christ's office was not to command but to persuade and counsel. So it is difficult to see how Christ could forbid anything. Moreover, Hobbes argued that (according to the bible) all we need to do in order to attain salvation is to believe in Christ and to obey the laws of nature and of the sovereign; the law of nature is silent on excommunicates; unless the sovereign has made a law on how we should act towards them, it is therefore hard to see what obligation we have to heed Christ's injunction. Marsilius of Padua 1956, 150 (2:6:13) similarly suggests that the biblical precepts on avoiding excommunicates are counsel not command and that they may be ignored with impunity.

47. Presbyterians commonly argued that the church meant the pastors and elders only: e.g. Herle 1643, 17; Bastwick 1624, 29. Independents standardly claimed that the church means all the members of a congregation (or at least all who were male heads of households): e.g. Jacob 1613, 322; Davenport 1636, 36. Almain 1606a, 720; Mair 1606, 878; cf. Sarpi 1607a, 613–14.

48. De Dominis 1620, 463: 'Non Communitati, sed Praelatis data est potestas iurisdictionis'. Hooker 1977–81, 3:386–7 (VIII, vi, 2; ecclesiastical power not held by the prelacy alone); 3:170–7 (VII, vi, 1–8; bishops).

49. Andrewes 1851b, 51; Thomson 1611, 58–60. Thomson states that the power of excommunication is held by the church, meaning not the mass of the people but the pastors, or more strictly the Apostles and their successors (i.e. the bishops): 'De Excommunicatione majore . . . . Ecclesiae ergo data potestas, non Petro (58) . . . Deinde scire licet R. Episcopum nomine Ecclesiae non plebem, sed pastores intelligere; (Neque enim existimat potestatem clavium a plebe in pastores derivari) & opponere multos uni. Id est, non soli Petro, & Petri successoribus, sed reliquis Apostolis, & eorum successoribus aequaliter datam esse potestatem illam' (59–60).

50. Marsilius of Padua 1956, 103–4, 149 (2:2:3, 2:6:12). Cranmer 1846, 117.

51. Chillingworth 1838a. Chillingworth's views on episcopacy are briefly discussed in Mason 1914, 154–6.

CHAPTER 6: GOD, RELIGION AND TOLERATION

1. Coke 1662, 25. Different versions of the thesis that Hobbes was a theist and Christian may be found in e.g. Glover 1965; Pocock 1973; Pacchi 1988; Letwin 1976; Eisenach 1982; Schneider 1974; Hood

1964; Halliday, Kenyon and Reeve 1983. Informative discussions of Hobbes' concept of God include Brown 1962; Hepburn 1972. Treatments of Hobbes' religious ideas include Ryan 1983; Damrosch 1979; Milner 1988; Johnston 1986, 164–84. Zagorin 1990 contains useful information on the context of some of Hobbes' religious ideas, including his use (in Lev 42: 343–4/271; 43: 414/331) of the biblical example of Naaman to vindicate outward obedience to a sovereign's irreligious commands (especially 32–3, 109–11, 138–40, 144–7, 327).

2. Similar passages include DC 2:21 Annot; DC 15:14; Hobbes' comments on Descartes' *Meditations* in Descartes 1984–5, 2:127; Lev 12: 77/53; Lev 31: 250/190.

3. Fuller references to relevant passages from Hobbes, and useful discussion of them, may be found in Brown 1962 and Hepburn 1972. Aristotle 1933, 86–95 (II.ii.1–13; 994a1–994b30). ST 1a, q.2. art.3, resp; Aquinas 1955–7, 1:85–8 (book 1, chapter 13, 1–8). Herbert of Cherbury 1937, 275–6. Grotius 1718, 4 (1, 2).

4. These and other relevant passages are discussed in Pacchi 1988, especially at 173–9. On 1 February 1641, Christopher Villiers wrote to Mersenne, asserting that 'the English philosopher' ('le philosophe Anglois'; i.e. Hobbes) maintained that we can have no conception of God and therefore cannot prove His existence: CM 10:453.

5. ST 1a, q.12, art.12, 1 and ad primum: 'simplex forma'; 'ratio ad formam simplicem pertingere non potest, ut sciat de ea quid est; potest tamen de ea cognoscere, ut sciat an est'. Cf. e.g. Laud 1847–60, 2:125–6; Calvin 1975, 1:238– 41 (II, ii, 18–21).

6. ST 1a, q.3, prologue and art.1, resp: 'nullum corpus movet non motum'; 1a, q.13, art.3, resp: 'Deum cognoscimus ex perfectionibus procedentibus in creaturas ab ipso; quae quidem perfectiones in Deo sunt secundum eminentiorem modum quam in creaturis'. Aquinas 1955–7, 1:96 (1:14:2). Copleston 1962, 66–8 ('negative way').

7. A parallel passage is in Bacon 1937, 66.

8. ST 2a2ae q.2, art. 4.

9. Fitzherbert 1615, 37–43. Aquinas, however, denied that reason is sufficient to show that kings must be subject to priests: *De regimine principum*, book 1, chapter 14, in Aquinas 1978, 76–7.

10. ST 2a2ae, q.1, art.5, resp.; q.9, art.1. Calvin 1975, 1:474–5 (III, ii, 7). Chillingworth 1838b, 1:116 (chapter 1, section 9).

11. The term Socinianism was used narrowly to denote the specific doctrines (including that Christ's nature is not divine) of Faustus Socinus and his followers, and much more broadly to signify the idea that reason has a large part to play in theological matters. Discussions of the relationship between Socinian ideas and those of Chillingworth, Falkland and John Hales are in McLachlan 1951, 57–87; Orr 1967, 81, 99; Weber 1940, 197–205; Elson 1948, 115–21; Trevor-Roper 1989, 186–92. A brief discussion of the relationship between Hobbes' religious ideas and Socinianism is in Geach 1981, 552–4.

12. Falkland 1651, 241, 245. Cheynell 1644, sig. B1a.

13. Calvin 1975, 1:475 (III, 2, 3).
14. Lucy 1673, 148.
15. Grotius 1647, 116, claims that according to Protestants in general there have been no prophets since the time of the Apostles. Hooker 1977–81, 2:322–3 (V, lxvi, 3) argues that not long after the Apostles' time miracles grew far less common than they had earlier been. Calvin 1975, 2:319 (IV, iii, 4), says that prophets no longer exist or are at least less conspicuous than they were in the time of the Apostles; and (2:637: IV, xix, 19) that the age of miracles 'immediately ceased' after the apostolic era. According to Richard Montagu, miracles had taken place in early times so 'that the world might beleeve' the Christian message; but they had since been replaced by Scripture, which alone was now to regulate faith and belief: Montagu 1624, preface, sig. A1b. An excellent discussion of attitudes towards prophecy and miracles is in Thomas 1973, 146–73.
16. Ferriby 1653, sig. B3b: 'most of our new lights are but old darknesses'; James Cranford's *imprimatur*, prefixed to Edwards 1646, warns against 'pretended New lights'; 'A godly exhortation' (1642), in Rollins 1923, 147: 'When some that cannot read or write/ Shall tell us of a new found light'. Hobbes inveighs against 'fanatics, the new lights of this age' ('Phanaticos huius nova lumina secli') in *Historia Ecclesiastica*, line 49 (LW5:350).
17. Sommerville 1990, 250–2. A good discussion of contemporary attitudes towards providence is Worden 1985, 55–99.
18. Pacchi 1988, 186n53.
19. Walker 1964; Hill 1972, 141–2.
20. Sweet 1617, 128–9. Morton 1596, 94 (quotation), 9–12 (minimalist views).
21. ST 2a2ae, q.2, arts.7–8; art.5, resp.
22. Chillingworth 1838b, 1:319–20 (chapter 3, section 13), quoting Luke 12:48; ibid., 1.320, quoting Hebrews 11:6; 1:366 (chapter 3, section 52). Clarendon 1759, first pagination, 29.
23. Hales 1765, 2:41, 44–5. Falkland 1651, 265, 263. Taylor 1647, second pagination 28, 26.
24. Johnson 1974, 114–15.
25. Hales 1765, 2:40, 3:163. Falkland 1651, 148.
26. Falkland 1651, sig. b4b–c1a.
27. The best recent survey of mid-seventeenth-century attitudes towards toleration is Worden 1984. Though it concentrates on a slightly later period, Goldie 1991b is an excellent discussion of attitudes which underlay intolerance. A standard work which helpfully summarises many relevant primary sources is Jordan 1932–40.
28. Goodwin 1645b, 22–3, 11–12. Robinson 1645, 42, 3.
29. Milton 1644, 35, 34; 1974, 320. Luke 1963, 583.
30. Rutherford 1649, 135, 139, 77, 83; 1644, 369, 372.
31. ST 1a2ae q.19, art.5–6. Parsons 1612, 542, 278; 1690, 32–3.
32. Rutherford 1649, 262. Cotton 1649, 7.
33. Rutherford 1649, 20.

34. Goodwin 1646, 16, 61. Walwyn 1646, 8. Cook 1647a, 4. Cf. e.g. *A declaration by congregationall societies*, 5.
35. Taylor 1647, 199. Falkland 1651, '217'=225–6. *M.S. to A.S*, 55–9; Walwyn 1646, 8–13; Richardson 1649, 28.
36. Goodwin 1646, 27–31; Falkland 1651, '224'=232, 243; Taylor 1647, 192–3.
37. Falkland 1651, 224; Taylor 1647, 190.
38. Goodwin 1645a, 36. Rutherford 1648, second pagination 93; 1646, first pagination 200–19.
39. Barlow 1609, 160. Hammond 1645, 8.
40. Stillingfleet 1661, 69. Sanderson 1636, 32; 1686b, 136–7.
41. Taylor 1647, 219.
42. Locke 1967, 124, 140; 1968, 103, 105, 159n35. Tuck 1990, 153–71, argues that the two authors took much the same line on toleration in the 1670s. Farr 1990, 188–91, contrasts the views of Hobbes and Locke, noting that Hobbes advised sovereigns to insist on uniformity of worship, and that Locke argued 'for the freedom of "indifferent" practices as well as belief'; Farr concludes that in Hobbes' work tolerationist strains 'exist alongside more virulent strains of intoleration' (190). Two informative articles about post-Restoration debates on toleration and comprehension, and their relationship to the views of Hobbes and Locke are Marshall 1985; Goldie 1983.
43. Hill 1971, 37–9.
44. Erasmus 1971, 153, 155.
45. Ibid, 157.
46. Erasmus 1971, 153, 155, 157. Shaw 1658, 31. Moulin 1610, 104. Abernethie 1638, 23.
47. Bacon 1974, 27–8. Herbert of Cherbury 1937, 333.
48. Thomas 1973, 74–5. Taylor 1647, 178, 180. Sir Kenelm Digby to Hobbes, Paris, 17 January 1637, in Tönnies 1936, 88. Gee 1624, 47–8. Harsnet 1599; Darrell's exorcisms and Harsnet's charges are discussed in Thomas 1973, 576–9.
49. Rudolph Gualter to Bishop Sandys, Zurich, 8 October 1573, in Robinson 1845, 238. Erastus 1589, 74; 1844, 159–60. Burgess 1643, sig. A2a. *The last warning to all the inhabitants of London*, 2. Nedham 1650, 55.
50. Canne 1657, sig. a5a. Hill 1972, 241–6 (radicals and reform of universities). Gardiner 1903, 2:322–3n2 (Barebones).
51. Quoted in Packer 1969, 179.

Bibliography

Unless otherwise stated, the place of publication is London.

1: BIBLIOGRAPHIES

The standard bibliography of Hobbes' writings is H. Macdonald and M. Hargreaves, *Thomas Hobbes: a bibliography*, 1952. A useful bibliography of writings about Hobbes is William Sacksteder, *Hobbes studies (1879–1979): a bibliography*, Bowling Green, Ohio, 1982.

2: PRIMARY SOURCES

(A) Works of Hobbes

The standard edition of Hobbes' works is Sir William Molesworth, ed., *The English Works of Thomas Hobbes* (11 vols, 1839–45) and *Thomae Hobbes . . . Opera philosophica quae Latine scripsit omnia* (5 vols, 1839–45). A more accurate edition is being issued by Oxford University Press, but so far only *De Cive: the Latin version* (edited by Howard Warrender, Oxford 1983) has appeared. *De Cive: the English version* (edited by Warrender, Oxford 1983) is not by Hobbes. The definitive edition of *Leviathan*, edited by Noel Malcolm, will be published in the Oxford series, but for the moment the best available text is that edited by Richard Tuck in the series Cambridge Texts in the History of Political Thought, published by Cambridge University Press, 1991. Important modern editions of works by Hobbes are *Behemoth or the Long Parliament*, edited by Ferdinand Tönnies, 1889, reprinted with an introduction by Stephen Holmes, Chicago 1990; *Thomas White's "De Mundo" examined*, translated by Harold Whitmore Jones, Bradford 1976; *A Dialogue between a Philosopher and a Student of the Common Laws of England*, edited by Joseph Cropsey, Chicago 1971; *The Elements of Law Natural and Politic*, edited by Ferdinand Tönnies 1889, second edition with a new introduction by M. M. Goldsmith, 1969. The abbreviations used in referring to the above works are listed in the List of Abbreviations. The following is a list of some letters and other brief writings by Hobbes:

De Beer, G. R. (1950), ed., 'Some letters of Thomas Hobbes', *Notes and Records of the Royal Society* 7 (1950), 195–206.
Historical Manuscripts Commission, *13th Report, Appendix, part I. The Manuscripts of His Grace the Duke of Portland preserved at Welbeck Abbey*, vol. 2, 1893, 124–30 (includes letters from Hobbes to Newcastle).
Memorable sayings of Mr Hobbes in his books and at the table, 1680 (Wing H. 2251A).

Mintz, Samuel I. (1968), 'Hobbes on the law of heresy: a new manuscript', *Journal of the History of Ideas* 29 (1968), 409–14.
Skinner, Quentin, 'Hobbes on sovereignty: an unknown discussion', *Political Studies* 13 (1965), 213–18.
Tönnies, Ferdinand (1889–90), 'Siebzehn Briefe des Thomas Hobbes an Samuel Sorbière', *Archiv für Geschichte der Philosophie* 3 (1889–90), 58–71, 192–232.
Tönnies, Ferdinand (1903–4), 'Hobbes-Analekten', *Archiv für Geschichte der Philosophie* 17 (1903–4), 291–317.
Tönnies, Ferdinand (1905–6), 'Hobbes-Analekten II', *Archiv für Geschichte der Philosophie* 19 (1905–6), 153–75.
Tönnies, Ferdinand (1936), 'Contributions à l'histoire de la pensée de Hobbes', *Archives de Philosophie* 12 (1936), cahier II, 73–98.
Wikelund, Philip R. (1969), '"Thus I passe my time in this place". An unpublished letter of Thomas Hobbes', *English Language Notes* 6 (1969), 263–8.
Zagorin, Perez (1978), 'Thomas Hobbes's departure from England in 1640: an unpublished letter', *Historical Journal* 21 (1978), 157–60.

(B) Other Works

Abernethie, Thomas (1638), *Abjuration of poperie*, Edinburgh 1638 (STC 72).
Adams, Thomas (1633), *A commentary or, exposition upon the divine second epistle generall, written by the blessed apostle St Peter*, 1633 (STC 108).
Almain, Jacques (1606a), *Libellus de auctoritate ecclesiae*, in Jean Gerson, *Opera*, ed. Edmond Richer, 4 vols, Paris 1606, vol. 1, columns 705–50.
—— (1606b), *Quaestio resumptiva*, in Jean Gerson, *Opera*, ed. Edmond Richer, 4 vols, Paris 1606, vol. 1, columns 687–704.
Ames, William (1639), *Conscience with the power and cases thereof*, Leyden and London 1639 (STC 552).
Ammianus Marcellinus (1940–52), *Rerum gestarum libri qui supersunt*, in *Ammianus Marcellinus with an English translation by John C. Rolfe*, 3 vols, Loeb Classical Library, 1940–52.
Andrewes, Lancelot (1851a), *Responsio ad Apologiam Cardinalis Bellarmini*, ed. J. Bliss, Oxford 1851.
—— (1851b), *Tortura Torti*, ed. J. Bliss, Oxford 1854.
—— (1854), *Two answers to Cardinal Perron*, ed. J. Bliss, Oxford 1854.
Aquinas, Thomas, *Summa theologiae*, many editions, including one with an English translation of varying value edited by Thomas Gilby and others, 61 vols, 1964–80.
—— (1955–7), *Summa contra Gentiles*, translated by Anton C. Pegis, 5 vols, Garden City, New York, 1955–7.
—— (1978), *Selected political writings*, ed. A. P. D'Entrèves, Oxford 1978.
Aristotle (1933), *The Metaphysics books I–IX with an English translation by Hugh Tredennick, M.A.*, Loeb Classical Library 1933.
—— (1934), *The Nichomachean Ethics with an English translation by M. Rackham M.A.*, Loeb Classical Library 1934.
Ascham, Anthony (1649), *Of the confusions and revolutions of goverments* [sic], 1649 (Wing A 3922).

Bibliography

Aubrey, John, *'Brief lives,' chiefly of contemporaries, set down by Aubrey, between the years 1669 & 1696*, edited by Andrew Clark, 2 vols, Oxford 1898.
—— (1972), *Brief lives*, ed. Oliver Lawson Dick, Harmondsworth 1972.
Azorius, Johannes (1602), *Institutionum moralium pars prima*, Cologne 1602.
—— (1606), *Institutionum moralium pars secunda*, Rome 1606.
Bacon, Sir Francis (1937), *Essays*, The World's Classics no. 24, Oxford 1937.
—— (1974), *The advancement of learning*, ed. Arthur Johnston, paperback, Oxford 1974.
Ball, William (1645), *Tractatus de jure regnandi & regni: or, the sphere of government*, 1645 (Wing B 597).
Barclay, William (1600), *De Regno et regali potestate*, Paris 1600.
—— (1611), *Of the Authoritie of the Pope*, issued as part of Richard Sheldon, *Certain general reasons, proving the lawfulness of the oath of allegiance*, 1611 (STC 22393).
Barlow, William (1609), *An answer to a Catholike English-Man (so by himself entituled)*, 1609 (STC 1446).
Barret, William (1612), *Ius Regis*, 1612 (STC 1501).
Bastwick, John (1624), *Elenchus religionis papisticae*, Leyden 1624.
Baxter, Richard (1696), *Reliquiae Baxterianae: or Mr. Richard Baxter's narrative of the most memorable passages of his life and times*, 1696 (Wing B 1370).
Bellarmine, Robert (1602–3), *Disputationum . . . de controversiis Christianae fidei*, 4 vols, Venice 1602–3.
Bennet, Robert (1649), *King Charles' triall iustified*, 1649 (Wing B 1886).
Bilson, Thomas (1593), *The perpetual government of Christes church*, 1593 (STC 3065).
Bodin, Jean (1606), *The six bookes of a commonweale*, translated by Richard Knolles, 1606 (STC 3193); there is a facsimile of this edition edited by Kenneth Douglas McRae, Cambridge Mass., 1962.
Bolton, Robert (1635), *Two sermons preached at Northampton*, 1635 (STC 3256).
Bramhall, John (1658), *The catching of Leviathan*, 1658, printed with *Castigations of Mr. Hobbes his last animadversions*, 1657 (Wing B 4214).
Bridge, William (1642), *The wounded conscience cured*, 1642 (Wing B 4476).
—— (1643), *The truth of the times vindicated*, 1643 (Wing B 4467).
Bridges, John (1587), *A defence of the government established*, 1587 (STC 3734).
Buckeridge, John (1614), *De potestate papae in rebus temporalibus*, 1614 (STC 4002).
Burgess, Cornelius (1643), *The necessity of agreement with God*, 1643 (Wing B 5673).
Burhill, Robert (1611), *Pro Tortura Torti*, 1611 (STC 4118).
—— (1613), *De potestate regia*, Oxford 1613 (STC 4117).
Burroughes, Jeremiah (1643), *The glorious name of God the Lord of hosts*, 1643 (Wing B 6074).
Burton, Thomas (1828), *Diary of Thomas Burton*, ed. J. T. Rutt, 4 vols, 1828.
Calvin, John (1975), *Institutes of the Christian religion*, 2 vols, translated by Henry Beveridge, Grand Rapids, Michigan, 1975.

Canne, John (1649), *The Golden Rule*, 1649 (Wing C 440).
—— (1657), *The time of the end*, 1657 (Wing C 443).
Carpenter, Nathaniel (1629), *Achitophel, or, the picture of a wicked politician*, 1629 (STC 4669).
Cartwright, Thomas (1573), *A replye to an answere made of M. Doctor Whitgift agaynste the admonition to the parliament*, [1573] (STC 4711).
Cary, Henry (1842), ed., *Memorials of the Great Civil War*, 2 vols, 1842.
Casaubon, Isaac (1607), *De libertate ecclesiastica*, Paris 1607.
Chaloner, Thomas (1646), *An answer to the Scotch papers*, 1646 (Wing C 1802).
Charles I (1628), *His Maiesties declaration to all his loving subiects, Of the causes which moved him to dissolve the last Parliament*, 1628 (STC 9249).
—— (1640), *His Majesties declaration: to all his loving subjects, Of the causes which moved him to dissolve the last Parliament*, 1640 (STC 9262).
Cheynell, Francis (1644), *Chillingworthi novissima, or, the sicknesse, heresy, death ... of William Chillingworth*, 1644 (Wing C 3810).
Chillingworth, William (1838a), *The apostolical institution of episcopacy demonstrated*, in *The Works of William Chillingworth*, M.A., 3 vols, Oxford 1838, 2:485–91.
—— (1838b), *The religion of Protestants a safe way to salvation*, in *The works of William Chillingworth M.A.*, 3 vols, Oxford 1838, vol. 1; vol. 2:1–484.
Cicero, Marcus Tullius (1913), *De officiis, with a translation by Walter Miller*, Loeb Classical Library, 1913.
Clarendon, Edward Hyde, Earl of (1676), *A brief view and survey of the dangerous and pernicious errors to church and state in Mr Hobbes book, entituled Leviathan*, Oxford 1676 (Wing C 4420).
—— (1759), *The life of Edward Earl of Clarendon*, Oxford 1759.
A cleere and full vindication of the late proceedings of the armie under the conduct of his excellencie, Sir Thomas Fairfax, 1647 (Wing C 4617).
Coke, Sir Edward (1658), *The Reports of Sir Edward Coke Kt. Late Lord Chief-Justice of England*, 1658 (Wing C 4944).
Coke, Roger (1662), *A survey of the politicks of Mr. Thomas White, Mr. Thomas Hobbs, and Mr. Hugo Grotius*, 1662 (Wing C 4982).
Coleman, Thomas (1645), *Hopes deferred and dashed*, 1645 (Wing C 5053).
—— (1646), *A brotherly examination re-examined*, 1646 (Wing C 5048).
Collins, Samuel (1617), *Epphata to F.T.*, Cambridge 1617 (STC 5561).
Constitutions and canons ecclesiasticall, 1640 (STC 10080).
Controversiae memorabilis inter Paulum V. Pontificem Max. & Venetos, Villa Sanvincentiana, 1607.
The Convocation book of MDCVI. Commonly called Bishop Overall's convocation book, Oxford 1844.
Cook, John (1647a), *Redintegratio amoris, or a union of hearts*, 1647 (Wing C 6026).
—— (1647b), *What the Independents would have*, 1647 (Wing C 6031).
—— (1649), *King Charls his case*, 1649 (Wing C 6025).
Cooke, James (1608), *Juridica trium quaestionum*, Oxford 1608 (STC 5671).
Cope, Esther S., and Coates, William H. (1977), eds, *Proceedings of the Short Parliament of 1640*, Camden Society, fourth series, vol. 19, 1977.

Cosin, Richard (1584), *An answer to the two first and principall treatises of a certeine factious libell*, 1584 (STC 5819.5).
Cotton, John (1645), *The way of the churches of Christ in New-England*, 1645 (Wing C 6471).
—— (1647), *The bloudy tenent, washed*, 1647 (Wing C 6409).
—— (1649), *The controversie concerning liberty of conscience*, 1649 (Wing C 6421).
Cowell, John (1607), *The Interpreter*, Cambridge 1607 (STC 5900).
Cranmer, Thomas (1846), 'Questions and answers concerning the sacraments and the appointment and power of bishops and priests', in *Miscellaneous writings and letters of Thomas Cranmer*, ed. J. E. Cox, Parker Society, Cambridge 1846, 115–17.
Danaeus, Lambertus (1577), *Ethices Christianae libri tres*, Geneva 1577.
Davenport, John (1636), *An apologeticall reply*, Rotterdam 1636 (STC 6310).
A declaration by congregationall societies, 1647 (Wing D 561).
De Dominis, Marc' Antonio (1617), *De republica ecclesiastica libri X. (Pars I. Continens libros I. II. III. IV . . .)*, 1617 (STC 6994).
—— (1620), *De republica ecclesiastica pars secunda: continens libros quintum et sextum*, 1620 (STC 6995.5).
Descartes, René (1964–76), *Oeuvres* ed. Charles Adam and Paul Tannery, 12 vols, Paris 1964–76.
—— (1984–5), *The philosophical writings of Descartes*, translated by John Cottingham, Robert Stoothoff and Dugald Murdoch, 2 vols., Cambridge 1984–5.
De Waard, Cornelis, and others, eds, *Correspondance du P. Marin Mersenne*, Paris, 17 vols, Paris 1932–88.
Diana, Antoninus (1660), *Practicae resolutiones*, Antwerp 1660.
Dickinson, William (1619), *The King's right; an assize sermon*, 1619 (STC 6821).
Digby, Sir Kenelm (1638), *A Conference with a lady about choice of religion*, Paris 1638 (STC 5625).
—— (1644), *Two treatises in the one of which, the nature of bodies; in the other, the nature of mans soule; is looked into*, Paris 1644 (STC 1448).
Digges, Dudley, (and perhaps others), (1642), *An answer to a printed book, intituled Observations upon some of his maiesties late answers and expresses*, Oxford 1642 (Wing D 1454).
—— (1643), *The unlawfulnesse of subjects taking up armes against their soveraigne*, Oxford 1643 (Wing D 1462).
Donne, John (1610), *Pseudo-Martyr*, 1610 (STC 7034).
—— (1984), *Biathanatos* (1608), ed. Ernest W. Sullivan II, Newark 1984.
Doughty, John (1651), *Velitationes polemicae: or, polemicall short discussions*, 1651 (Wing D 1963).
Dove, John (1610), *An advertisement to the English Iesuites and seminaries*, 1610 (STC 7077).
Downing, Calybute (1632), *A Discourse of the state ecclesiasticall*, Oxford 1632 (STC 7156).
Eachard, John (1958), *Mr. Hobbs's state of nature considered in a dialogue between Philautus and Timothy* (1672), edited by Peter Ure, English Reprint series no. 14, Liverpool 1958.

—— (1673), *Some opinions of Mr. Hobbes considered*, 1673 (Wing E 64).
Edwards, Thomas (1646), *Gangraena*, 1646 (Wing E 228).
Eliot, Sir John (1879), *The Monarchie of Man*, ed. A. B. Grosart, 2 vols 1879.
Ellesmere, Thomas Egerton, Baron (1977), *Ellesmere's Tracts*, in Louis A. Knafla, *Law and Politics in Jacobean England*, Cambridge 1977, 195–336.
Erasmus, Desiderius (1971), *Praise of Folly*, translated by Betty Radice, Harmondsworth 1971.
Erastus, Thomas (1589), *Explicatio gravissimae quaestionis*, 'Pesclavii' (i.e. London) 1589 (STC 10511).
—— (1844), *The Theses of Erastus touching excommunication*, translated and edited by Robert Lee, Edinburgh 1844.
Falkland, Lucius Cary, Viscount (1651), *Sir Lucius Cary, Late Lord Viscount Falkland, his Discourse of Infallibility, with an answer to it: and his lordships reply, 1651* (Wing F 317).
Ferne, Henry (1643a), *Conscience satisfied*, Oxford 1643 (Wing F 791).
—— (1643b), *A reply unto severall treatises*, 1643 (Wing F 799).
Ferriby, John (1653), *The lawfull preacher: or a short discourse: proving that they only ought to preach who are ordained ministers*, 1653 (Wing F 819A).
Field, Richard (1847–52), *Of the church, five books*, 4 vols, Cambridge 1847–52.
Filmer, Sir Robert (1991), *Patriarcha and other writings*, ed. Johann P. Sommerville, Cambridge 1991.
Finch, Sir Henry (1627), *Law, or, a discourse thereof*, 1627 (STC 10871).
Firth, C. H. (1895), 'A letter from Lord Saye and Sele to Lord Wharton, 26 December 1657', *English Historical Review* 10 (1895), 106–7.
Firth, C. H., and R. S. Rait, eds (1911), *Acts and Ordinances of the Interregnum, 1642–1660*, 3 vols, 1911.
Fitzherbert, Thomas (1613), *A supplement to the discussion of M. D. Barlowes answere*, St Omer 1613 (STC 11021).
—— (1615), *The second part of a treatise concerning policy and religion*, Douai 1615 (STC 11019.5).
Floyd, John (1620), *God and the King. Or a dialogue wherein is treated of Allegiance due to our most gracious lord, King Iames*, Cologne [really St Omer] 1620 (STC 11110.7).
Fortescue, Sir John (1616), *De laudibus legum Angliae*, ed. John Selden, 1616 (STC 11197).
Foster, Elizabeth Read, ed. (1966), *Proceedings in Parliament 1610*, 2 vols, New Haven 1966.
Fulbecke, William (1602), *The pandectes of the law of nations*, 1602 (STC 11414).
—— (1618), *A Parallele or conference of the civil law, the canon law, and the common law*, 1618 (STC 11416).
Fulke, William, ed. (1601), *The text of the New Testament of Iesus Christ, Translated out of the vulgar Latine by the papists of the traiterous seminarie at Rhemes*, 1601 (STC 2900).
Fussner, Frank Smith (1957), 'William Camden's "Discourse concerning the prerogative of the Crown', in *Proceedings of the American Philosophical Society* 101 (1957), 204–15.
Gaius (1940), 'Gai Institutionum commentarii quattuor', in J. Baviera, ed.,

Fontes iuris Romani antejustiniani. Pars altera: auctores, Florence 1940, 1–192.
Gassendi, Pierre (1963), *Dissertations en forme de paradoxes contre les Aristoteliciens (Exercitationes paradoxiae adversus Aristoteleos), livres I et II*, translated by Bernard Rochot, Paris 1963.
Gee, Edward (1658), *The divine right and originall of the civil magistrate*, 1658 (Wing G 448).
Gee, John (1624), *The foot out of the snare*, 1624 (STC 11701).
Gerson, Jean (1606a), *Descriptiones terminorum ad theologiam utilium*, in *Opera*, ed. Edmond Richer, 4 vols, Paris 1606, vol.4, columns 125–44.
—— (1606b), *Liber de vita spirituali animae*, in *Opera*, ed. Edmond Richer, 4 vols, Paris 1606, vol.3, columns 160–239.
Gillespie, George (1645), *A sermon preached before the right honourable the House of Lords, in the abbey church at Westminster, upon the 27th. of August. 1645*, 1645 (Wing G 758).
Goodwin, John (1645a), *Calumny arraigned and cast*, 1645 (Wing G 1153).
—— (1645b), *Innocency and truth triumphing together*, 1645 (Wing G 1176).
—— (1646), *Hagiomastix, or the scourge of the saints displayed*, 1646 (Wing G 1169).
Gordon, John (1610), *Antitortobellarminus*, 1610 (STC 12054).
Grotius, Hugo (1647), *De imperio summarum potestatum circa sacra*, Paris 1647.
—— (1689), *De jure belli ac pacis*, Amsterdam 1689.
—— (1718), *De veritate religionis Christianae*, Amsterdam 1718.
Hale, Sir Matthew (1924), 'Reflections by the Lrd. Cheife Justice Hale on Mr. Hobbes his dialogue of the lawe', in W. S. Holdsworth, *A history of English law*, vol. 5, Boston, Mass., 1924, 500–513.
Hales, John (1765), *The works of the ever memorable Mr. John Hales of Eaton*, 3 vols, Glasgow 1765.
Hammond, Henry (1645), *Of conscience*, 1645 (Wing H 549).
—— (1650) *A vindication of Dr Hammond's addresse*, 1650 (Wing H 616).
Harrington, James (1977), *The political works of James Harrington*, ed. J. G. A. Pocock, Cambridge 1977.
Harris, Richard (1614), *The English Concord, In answere to Becanes English Iarre*, 1614 (STC 12815).
Harsnet, Samuel (1599), *A discovery of the fraudulent practises of John Darrel*, 1599 (STC 12883).
Hayward, Sir John (1603), *An answer to the first part of a certain conference, concerning succession*, 1603 (STC 12988).
—— (1606), *A reporte of a discourse concerning supreme power in affaires of religion*, 1606 (STC 13001).
Herbert of Cherbury, Edward, Lord (1937), *De Veritate*, translated Meyrick H. Carré, Bristol 1937.
Herle, Charles (1642), *A fuller answer to a treatise written by Doctor Ferne*, 1642 (Wing H 1558).
—— (1643) *The independency on Scriptures of the independency of churches*, 1643 (Wing H 1559).
Heylin, Peter (1637), *A briefe and moderate answer, to the seditious and scandalous challenges of Henry Burton*, 1637 (STC 13269).

—— (1658), *The stumbling-block of disobedience and rebellion*, 1658 (Wing H 1736).
Hooker, Richard (1977–81), *Of the lawes of ecclesiasticall politie*, in *The Folger edition of the works of Richard Hooker*, vols 1–3, Cambridge Mass., 1977–81.
Hooker, Thomas (1648), *A survey of the summe of church-discipline*, 1648 (Wing H 2658).
Hudson, Michael (1647), *The divine right of government*, 1647 (Wing H 3261).
'Illustrations of the state of the church during the Great Rebellion', *The Theologian and Ecclesiastic* 6 (1848), 161–75.
Isham, Gyles, Bart., ed. (1951), *The Correspondence of Bishop Duppa and Sir Justinian Isham 1650–1660*, Northamptonshire Record Society, vol. 17, 1951.
Jackson, Thomas (1844), *A treatise of Christian obedience*, in *Works*, 12 vols, Oxford 1844, vol. 12.
Jacob, Henry (1613), *An attestation of many learned . . . divines that the church-government ought to bee alwayes with the peoples free consent*, Middelburg 1613 (STC 14328).
James I (1621), *His Maiesties Declaration, touching his proceedings in the late assemblie and convention of Parliament*, 1621 (STC 9241).
—— (1918), *The political works of James I*, ed. C. H. McIlwain, Cambridge, Mass, 1918.
Jansson, Maija (1988), ed., *Proceedings in Parliament 1614 (House of Commons)*, Memoirs of the American Philosophical Society, vol. 172, Philadelphia 1988.
Johnson, R. C., Keeler, M. F., and others, eds (1977–83), *Proceedings in Parliament 1628*, 6 vols., New Haven 1977–83.
Josselin, Ralph (1976), *The diary of Ralph Josselin 1616–1683*, ed. Alan McFarlane, 1976.
Justinian, *Imperatoris Iustiniani Institutionum libri quattuor*, ed. J. B. Moyle, fifth edition, Oxford 1912.
Kellison, Matthew (1621), *The right and iurisdiction of the prelate, and the prince*, second edition, Douai 1621 (STC 14911).
King, John (1607), *The fourth sermon preached at Hampton Court*, 1607 (STC 14974).
Kynaston, Sir Francis (1629), 'A true presentation of forepast parliaments', B. L. Lansdowne MSS 213, ff. 146a–176b. Dated 1629 on the title-page.
The last warning to all the inhabitants of London, 1646 (Wing L 512).
Laud, William (1847–60), *Works*, ed. W. Scott and J. Bliss, 7 vols, Oxford 1847–60.
Lawson, George (1657), *An examination of the political part of Mr. Hobbs his Leviathan*, 1657 (Wing L 706).
Laymann, Paul (1643), *Theologiae moralis in V. lib. partitae*, Lyons 1643. First published in 1625..
Le Bret, Cardin (1632), *De la sovveraineté du roy*, Paris 1632.
Leibniz, Gottfried Wilhelm (1972), *The political writings of Leibnitz*, ed. Patrick Riley, Cambridge 1972.
—— (1985), 'Reflections on the work that Mr. Hobbes published in English

on "Freedom, Necessity and Chance', in Leibniz, *Theodicy*, translated by E. M. Huggard, ed. Austin Farrer, La Salle 1985, 393–404.
Lessius, Leonardus (1612), *De justitia et iure*, Antwerp 1612.
Locke, John (1967), *Two tracts on government*, ed. Philip Abrams, Cambridge 1967.
—— (1968), *Epistola de tolerantia: a letter on toleration*, ed. Raymond Klibansky and J. W. Gough, Oxford 1968.
—— (1988), *Two Treatises of government*, ed. Peter Laslett, Cambridge 1988.
Lucy, William (1673), *An answer to Mr. Hobbs his Leviathan*, 1673 (Wing L 3452).
Luke, Sir Samuel (1963), *The Letter Books of Sir Samuel Luke 1644–45*, ed. H. G. Tibutt, 1963.
M.S. to A.S. *With a plea for libertie of conscience*, 1644 (Wing S 116A).
Mair, John (1606), *Disputatio Ioannis Maioris doctoris theologi Parisiensis, de auctoritate concilii supra pontificem maximum*, in Jean Gerson, *Opera*, ed. Edmond Richer, 4 vols, Paris 1606, vol. 1, columns 875–94.
Maltby, Judith D., ed. (1988), *The Short Parliament (1640) Diary of Sir Thomas Aston*, Camden Society 1988.
March, John (1642), *An argument or, debate in law: of that great question concerning the militia*, 1642 (Wing M 575).
Marsilius of Padua (1956), *The defender of the peace, (Defensor pacis)*, translated by Alan Gewirth, New York 1956.
Maximes unfolded, 1643 (Wing M 1375).
Maynwaring, Roger (1627), *Religion and alegiance: in two sermons*, 1627 (STC 17751.5).
Milton, John (1644), *Areopagitica* (Wing M 2092).
—— (1974), *A treatise of civil power in ecclesiastical causes* (1659), in *Selected prose*, ed. C. A. Patrides, Harmondsworth 1974, 296–326.
Mitchell, A. F., and Struthers, J. (1874), eds, *Minutes of the sessions of the Westminster Assembly of divines*, 1874.
Montagu, Richard (1624), *A gagg for the new Gospell? No: a new gagg for an old goose*, 1624 (STC 18038).
Morton, Thomas, of Berwick (1596), 'Of the church', in *Salomon or a treatise declaring the state of the kingdome of Israel, as it was in the daies of Salomon, Whereunto is annexed another treatise Of the Church*, 1596 (STC 18197.3).
Morton, Thomas, Bishop (1610), *The encounter against M. Parsons*, 1610 (STC 18183).
Moulin, Lewis Du (1650), *The power of the Christian magistrate in sacred things*, 1650 (Wing D 2551).
Moulin, Pierre Du, the elder (1610), *A defence of the catholicke faith*, 1610 (STC 7322).
—— (1614), *De Monarchia temporali pontificis Romani*, 1614 (STC 7335).
Moulin, Pierre Du, the younger (1640), *A letter of a French Protestant to a Scotishman of the Covenant*, 1640 (STC 7345).
Nedham, Marchamont (1650), *The case of the common-wealth of England stated*, second edition, 1650 (Wing N 376).
—— (1659), *Interest will not lie*, 1659 (Wing N 392).

Nicastro, Onofrio (1973), *Lettere di Henry Stubbe a Thomas Hobbes*, Siena 1973.
Notestein, Wallace, Relf, Frances Helen, and Simpson, Hartley, eds (1935), *Commons Debates 1621*, 7 vols, New Haven 1935.
Ogle, O., Bliss, W. H., and Macray, W. D., eds (1869–76), *Calendar of the Clarendon State Papers*, 3 vols, Oxford 1869–76.
Owen, David (1622), *Anti-Paraeus: sive determinatio de iure regio*, Cambridge 1622 (STC 18982).
Parker, Henry (1640), *The case of shipmony briefly discoursed*, 1640 (STC 19215).
—— (1642), *Observations upon some of his Majesties late answers and expresses*, second edition, 1642 (Wing P 413).
—— (1644), *Jus Populi*, 1644 (Wing P 403).
Parsons, Robert (1606), *An answere to the fifth part of reportes lately set forth by Syr Edward Cooke*, St Omer 1606 (STC 19352).
—— (1612), *A Discussion of the answere of M. William Barlow*, 1612 (STC 19409).
—— (1690), *The Jesuit's Memorial for the intended reformation of England*, 1690 (Wing P 569).
Pemble, William (1632), *A summe of morall philosophy*, Oxford 1632 (STC 19587).
Philodemius, Eutactus (1649), *The original & end of civil power*, 1649 (Wing A 3921).
—— (1650), *An answer to the vindication of Doctor Hammond, against the exceptions of Eutactus Philodemius*, 1650 (Wing A 3918).
Preston, Thomas, (alias Roger Widdrington) (1611), *Apologia Cardinalis Bellarmini pro iure principum*, 1611 (STC 25596).
—— (1612), *Responsio apologetica ad libellum cuiusdam Doctoris Theologi*, 1612 (STC 25597).
—— (1619), *Last Reioynder to Mr. Thomas Fitzherberts Reply*, 1619 (STC 25599).
Prynne, William (1642), *A soveraigne antidote to prevent, appease, and determine our unnaturall and destructive civill wars*, second edition, 1642 (Wing P 4087).
—— (1645), *A vindication of foure serious questions*, 1645 (Wing P 4124).
Pym, John (1641), *The speech or declaration of John Pym*, in *Speeches and passages of this great and happy Parliament*, 1641 (Wing E 2309), sig. 3E1a-3F3b.
Richardson, Samuel (1649), *An answer to the London ministers letter*, 1649 (Wing R 1402).
Richer, Edmond (1692), *De potestate ecclesiae in rebus temporalibus*, Cologne 1692.
Ridley, Thomas (1634), *A view of the civile and ecclesiasticall law*, second edition, Oxford 1634 (STC 21055).
Robinson, H., ed. (1845), *The Zurich letters*, second series, Cambridge, Parker Society, 1845.
Robinson, Henry (1645), *A moderate answer to Mr. Prins full reply*, 1645 (Wing R 1676).
Rogers, Thomas (1607), *The Faith, Doctrine, and religion, professed, & protected*

in the Realme of England, Cambridge 1607 (STC 21228).
Rollins, Hyder E., ed. (1923), *Cavalier and puritan,* New York 1923.
Rushworth, John, ed. (1659–1701), *Historical Collections,* 7 vols., 1659–1701.
Rutherford, Samuel (1644), *The due right of presbyteries,* 1644 (Wing R 2378).
—— (1646), *The divine right of church-government and excommunication,* 1646 (Wing R 2377).
—— (1648), *A survey of the spiritual Antichrist,* 1648 (Wing R 2394).
—— (1649), *A free disputation against pretended liberty of conscience,* 1649 (Wing R 2379).
—— (1843), *Lex Rex,* Edinburgh 1843.
St German, Christopher (1974), *St German's Doctor and Student,* ed. T. F. T. Plucknett and J. L. Barton, Selden Society 1974.
Sanchez, Thomas (1654), *De sancto matrimonii sacramento,* Lyons 1654. First published 1602–5.
Sanderson, Robert (1636), *Two sermons,* 1636 (STC 21710).
—— (1661), *Episcopacy (as established by law in England) not prejudicial to regal power,* 1661 (Wing S 599).
—— (1674), *Eight Cases of conscience: occasionally determined,* 1674 (Wing S 598).
—— (1686a), *De juramenti promissorii obligatione praelectiones septem,* 1686 (Wing S 587).
—— (1686b), *De obligatione conscientiae praelectiones decem,* 1686 (Wing S 595).
Saravia, Hadrian (1592), *Of the diverse degrees of the ministers of the gospell,* 1592 (STC 21750).
—— (1611), *De imperandi authoritate,* in *Diversi tractatus theologici,* 1611 (STC 21751), fourth pagination, 107–314.
Sarpi, Paolo (1607a), *Apologia adversus oppositiones factas ab illustrissimo & reverendiss. Domino Card. Bellarmino ad tractatus, & resolutiones Iohan. Gersonis circa valorem excommunicationum,* in *Controversiae memorabilis inter Paulum V. Pontificem Max. & Venetos,* Villa Sanvincentiana, 1607, second pagination, 477–662.
—— (and others) (1607b), *Tractatus de interdicto sanctitatis papae Pauli V,* in *Controversiae memorabilis inter Paulum V. Pontificem Max. & Venetos,* Villa Sanvincentiana, 1607, first pagination, 169–242.
Sclater, William (1616), *A sermon preached at the last general assise holden for the county of Sommerset,* 1616 (STC 21843).
Selden, John (1616), 'Notes upon Fortescue', in Sir John Fortescue, *De laudibus legum Angliae,* 1616 (STC 11197).
—— (1618), *The historie of tithes,* 1618 (STC 22172).
—— (1640), *De jure naturali & gentium, iuxta disciplinam Ebraeorum, libri septem,* 1640 (STC 22168).
—— (1650), *De synedriis & praefecturis iuridicis veterum Ebraeorum. Liber primus,* 1650 (Wing S 2425).
—— (1652), *Of the dominion, or ownership of the sea,* translated by Marchamont Nedham, 1652 (Wing S 2432).
—— (1927), *The table talk of John Selden,* ed. Sir Frederick Pollock, 1927.
Shaw, Samuel, (1658), *Holy things for Holy men,* 1658 (Wing S 3037).

Sheldon, Richard (1611), *Certain general reasons, proving the lawfulness of the oath of allegiance*, 1611 (STC 22393).
Sibthorp, Robert (1627), *Apostolike Obedience*, 1627 (STC 22525.5).
Sidney, Algernon (1990), *Discourses concerning government*, edited by Thomas G. West, Indianapolis 1990.
Smith, Richard (1631), *A conference of the Catholike and Protestant doctrine with the expresse words of Holie Scripture*, Douai 1631 (STC 22810).
Speeches and passages of this great and happy Parliament, 1641 (Wing E 2309).
Spelman, Sir John (1642), *A view of a printed book intituled Observations upon his Majesties late answers and expresses*, 1642 (Wing S 4941).
Spelman, Sir John (judge) (1978), *The reports of Sir John Spelman*, ed. J. H. Baker, 2 vols, Selden Society 1978.
Stillingfleet, Edward (1661), *Irenicum. A weapon-salve for the churches wounds*, 1661 (Wing S 5596).
Suarez, Francisco (1856–78), *Opera*, 27 vols, Paris 1856–78 (vols. 5–6: *De legibus*; vol. 24: *Defensio fidei Catholicae*).
—— (1971–81), *De legibus*, ed. Luciano Pereña and others, 8 vols, Madrid 1971–81 (Corpus Hispanorum De Pace).
—— (1973), 'Additiones Suarecii ad ius gentium', in Suarez, *De legibus (II 13–20) Dé iure gentium*, ed. L. Pereña and others, Madrid 1973 (Corpus Hispanorum De Pace XIV), 150–165.
—— (1978), *De iuramento fidelitatis* (book 6 of *Defensio fidei Catholicae*) ed. L. Pereña et al, Madrid 1978 (Corpus Hispanorum De Pace XIX).
Sutcliffe, Matthew (1591), *De presbyterio*, 1591 (STC 23458).
Sweet, John (1617), *Monsigr. fate voi. Or a discovery of the Dalmatian Apostata*, St Omer 1617 (STC 23529).
Symons, Henry (1658), *A beautiful swan with two black feet*, 1658 (Wing S 6360B).
Taylor, Jeremy (1647), *A discourse of the liberty of prophesying*, 1647 (Wing T 400).
—— (1851), *Ductor dubitantium*, books 1 and 2, in *The whole works of the Right Rev. Jeremy Taylor, D.D.*, ed. Reginald Heber and others, 10 vols, vol. 9, 1851. References are to book, chapter, rule and section.
—— (1864), *Ductor dubitantium*, books 3 and 4, in ibid., vol. 10, 1864. References are to book, chapter, rule and section.
Templer, John (1673), *Idea Theologiae Leviathanis*, 1673 (Wing T 664).
Tenison, Thomas (1670), *The creed of Mr Hobbes*, 1670 (Wing T 691).
Thomson, Richard (1611), *Elenchus refutationis Torturae Torti*, 1611 (STC 24032).
Tönnies, Ferdinand (1936), 'Contributions à l'histoire de la pensée de Hobbes', *Archives de Philosophie* 12(1936), cahier II, 73–98.
Tooker, William (1611), *Duellum sive singulare certamen cum Martino Becano Iesuita*, 1611 (STC 24119).
Towerson, Gabriel (1676), *An explication of the decalogue or ten commandments*, 1676 (Wing T 1970).
Vaughan, William (1608), *The Golden-grove, moralized in three books*, second edition, 1608 (STC 24611).
Vedel, Nicolaus (1647), *The supreme power of Christian states vindicated against the insolent pretences of Guillielmus Apollonii*, 1647 (Wing V168).

Vox militaris: or an apologeticall declaration concerning the officers and souldiers of the army, 1647 (Wing V 721).
Vox plebis, 1646 (Wing V 726).
Walker, Clement (1648), *The History of Independency*, 1648 (Wing W 329).
Walwyn, William (1646), *Tolleration iustified and persecution condemned*, 1646 (Wing W 692A).
Ward, Seth, and Wilkins, John (1654), *Vindiciae Academiarum*, Oxford 1654 (Wing W 832).
Warmington, William (1612), *A moderate defence of the oath of allegiance*, 1612 (STC 25076).
Warner, G. F., ed. (1886), *Correspondence of Sir Edward Nicholas*, Camden Society, new series, 40(1886).
Watson, Richard (1651), [Greek: Akoulouthos] *or a second fair warning to take heed of the Scottish discipline*, The Hague, 1651 (Wing W 1084).
Whitehall, John (1679), *The Leviathan found out*, 1679 (Wing W 1866).
—— (1680), *Behemoth arraign'd: or, a vindication of property against a fanatical pamphlet stiled Behemoth*, 1680 (Wing W 1865).
Wilson, Matthew (1638), *Motives maintained. Or a reply to M. Chillingworthes answere*, St Omer 1638 (STC 25780).
Wood, Anthony à (1813–20), *Fasti Oxonienses or Annals of the University of Oxford . . . the first part*, in *Athenae Oxonienses*, ed. Philip Bliss, 4 vols, 1813–20, vol. 2.
Woodhouse, A. S. P. (1938), *Puritanism and Liberty Being the army debates (1647–9) from the Clarke manuscripts*, 1938.

3: SECONDARY SOURCES

Abbott, Philip (1981), 'The three families of Thomas Hobbes', *The Review of Politics*, 43(1981), 242–58.
Allen, J. W. (1928), *A history of political thought in the sixteenth century*, 1928.
Ashcraft, Richard (1971), 'Hobbes's natural man: a study in ideology formation', *Journal of Politics* 33(1971) 1076–1171.
Aylmer, G. E., ed. (1972), *The Interregnum: the quest for settlement 1646–1660*, 1972.
Barnouw, Jeffrey (1980), 'Hobbes's causal account of sensation', *Journal of the History of Philosophy* 17(1980), 115–30.
Barry, Brian (1972), 'Warrender and his critics', in Maurice Cranston and Richard S. Peters, eds, *Hobbes and Rousseau: a collection of critical essays*, Garden City, New York, 1972, 37–65.
Baumgold, Deborah (1983), 'Subjects and soldiers: Hobbes on military service', in *History of Political Thought* 4(1983), 43–64.
—— (1988), *Hobbes's political theory*, Cambridge 1988.
Bedford, R. D. (1979), *The defence of truth: Herbert of Cherbury and the seventeenth century*, Manchester 1979.
Bowle, John (1951), *Hobbes and his critics*, 1951.
Brandt, Frithiof (1928), *Thomas Hobbes' mechanical conception of nature*, Copenhagen 1928.

Brown, Keith C. (1962), 'Hobbes' grounds for belief in a deity', in *Philosophy* 37(1962), 336–44.
—— ed. (1965), *Hobbes Studies*, Cambridge Mass, 1965.
Burgess, G. (1985), 'Usurpation, obligation and obedience in the thought of the Engagement controversy', *Historical Journal* 29(1985), 27–50.
Burn, Richard (1788), *Ecclesiastical Law*, fifth edition, 4 vols 1788.
Burns, J. H., ed. (1988), *The Cambridge History of Medieval Political Thought c.350–c.1450*, Cambridge 1988.
—— ed. (1991), *The Cambridge History of Political Thought, 1450–1700*, Cambridge 1991.
Chapman, Richard Allen (1975), '*Leviathan* writ small: Thomas Hobbes on the family', in *American Political Science Review* 69(1975), 76–90.
Church, William Farr (1941), *Constitutional thought in sixteenth-century France: a study in the evolution of ideas*, Cambridge, Mass., 1941.
Coleman, Janet (1988), 'Property and poverty', in J. H. Burns, ed., *The Cambridge History of Medieval Political Thought c.350– c.1450*, Cambridge 1988.
Collinson, Patrick (1979), 'If Constantine, then also Theodosius: St Ambrose and the integrity of the *Ecclesia Anglicana*', in *Journal of Ecclesiastical History* 20(1979), 205–29.
Copleston, Frederick (1962), *A history of philosophy, volume 2: Medieval Philosophy, part II: Albert the Great to Duns Scotus*, paperback edition, Garden City, New York, 1962.
—— (1964), *A history of philosophy: volume 5: modern philosophy: the British philosophers part I: Hobbes to Paley*, paperback edition, Garden City, New York 1964.
Cranston, Maurice, and Peters, Richard S., eds (1972), *Hobbes and Rousseau: a collection of critical essays*, New York 1972.
Cust, Richard (1987), *The Forced Loan and English Politics, 1626–1628*, Oxford 1987.
Daly, James (1978), 'The idea of absolute monarchy in seventeenth-century England', *Historical Journal* 21(1978), 227–50.
—— (1979), *Sir Robert Filmer and English political thought*, Toronto 1979.
Damrosch, Leopold, Jr. (1979), 'Hobbes as Reformation theologian: implications of the freewill controversy', in *Journal of the History of Ideas* 40(1979), 339–52.
D'Entrèves, A. P. (1970), *Natural Law*, 1970.
Dietz, Mary G., ed. (1990), *Thomas Hobbes and political theory*, Lawrence, Kansas, 1990.
Du Boulay, F. R. H. (1970), *An age of ambition: English society in the late Middle Ages*, New York 1970.
Dzelzainis, Martin (1989), 'Edward Hyde and Thomas Hobbes's *Elements of Law, Natural and Politic*', in *Historical Journal* 32(1989), 303–17.
Eisenach, Eldon J. (1982), 'Hobbes on church, state and religion', in *History of Political Thought* 3(1982), 215–43.
Elson, James Hinsdale (1948), *John Hales of Eton*, New York 1948.
Farr, James (1990), 'Atomes of Scripture: Hobbes and the politics of biblical interpretation', in Mary Dietz, ed., *Thomas Hobbes and political theory*, Lawrence, Kansas, 1990, 172–196.

Feingold, Mordechai (1985), 'A friend of Hobbes and an early translator of Galileo: Robert Payne of Oxford', in J. D. North and J. J. Roche, eds, *The light of nature* (Dordrecht 1985), 265–80.

Figgis, John Neville (1914), *The Divine Right of Kings*, second edition, Cambridge 1914.

Gabrieli, Vittorio (1957), 'Bacone, la riforma e Roma nella versione Hobbesiana d'un carteggio di Fulgenzio Micanzio', *English Miscellany* 8(1957), 195–250.

Gardiner, S. R. (1883), *History of England from the Accession of James I to the outbreak of the Civil War, 1603–1642*, 10 vols, 1883.

—— (1903), *History of the Commonwealth and Protectorate 1649–1656*, 4 vols, 1903.

Gauthier, David (1969), *The logic of Leviathan*, Oxford 1969.

—— (1988), 'Hobbes's social contract', in G. A. J. Rogers and Alan Ryan, eds, *Perspectives on Thomas Hobbes*, Oxford 1988, 125–52.

Geach, P. T. (1981), 'The religion of Thomas Hobbes', in *Religious Studies* 17(1981), 549–58.

Glover, Willis B. (1965), 'God and Thomas Hobbes', in Keith C. Brown, *Hobbes Studies*, Cambridge Mass., 1965, 141–168.

Goldie, Mark (1983), 'John Locke and Anglican Royalism', in *Political Studies* 31(1983), 86–102.

—— (1991a), 'The reception of Hobbes', in J. H. Burns, ed., *The Cambridge History of Political Thought, 1450–1700*, Cambridge 1991, 589–615.

—— (1991b), 'The theory of religious intolerance in Restoration England', in Ole Peter Grell, Jonathan I. Israel, and Nicholas Tyacke, eds, *From persecution to toleration: the Glorious Revolution and religion in England*, Oxford 1991, 331–68.

Goldsmith, Maurice (1966), *Hobbes's science of politics*, New York 1966.

Greenleaf, W. H. (1974), 'A note on Hobbes and the Book of Job', *Annales de la Catedra Francisco Suarez* 14(1974).

Grover, Robinson A. (1980), 'The legal origin of Thomas Hobbes's doctrine of contract', in *Journal of the History of Philosophy* 18(1980), 177–194.

Gunn, J. A. W. (1969), *Politics and the public interest in the seventeenth century*, 1969.

Guth, D. J, and McKenna, J. W., eds (1982), *Tudor rule and revolution: essays for G. R. Elton from his American friends*, Cambridge 1982.

Halliday, R. J., Kenyon, Timothy, and Reeve, Andrew (1983), 'Hobbes's belief in God', in *Political Studies* 31(1983), 418–33.

Hamilton, James Jay (1978), 'Hobbes's Study and the Hardwick Library', in *History of Philosophy*, 16(1978), 445–53.

Hampton, Jean (1986), *Hobbes and the social contract tradition*, Cambridge 1986.

Heinze, R. W. (1982), 'Proclamations and parliamentary protest, 1539–1610', in D. J. Guth and J. W. McKenna, eds, *Tudor rule and revolution: essays for G. R. Elton from his American friends*, Cambridge 1982, 237–59.

Hepburn, R. W. (1972), 'Hobbes on the knowledge of God', in M. Cranston and R. S. Peters, eds, *Hobbes and Rousseau: a collection of critical essays*, Garden City, New York, 1972, 85–108.

Hill, Christopher (1971), *Antichrist in seventeenth-century England*, 1971.

—— (1972), *The world turned upside down*, 1972.
Hood, F. C. (1964), *The divine politics of Thomas Hobbes*, Oxford 1964.
Jacquot, Jean (1949–50), 'Un amateur de science, ami de Hobbes et de Descartes, Sir Charles Cavendish (1591–1654)', *Thales* 6(1949–50), 81–8.
Johnson, Paul J. (1974), 'Hobbes's Anglican doctrine of salvation', in Ross, Schneider and Waldman, eds, *Thomas Hobbes in his time*, Minneapolis 1974, 102–25.
Johnston, David (1986), *The rhetoric of Leviathan: Thomas Hobbes and the politics of cultural transformation*, Princeton 1986.
Jordan, W. K. (1932–40), *The development of religious toleration in England*, 4 vols, Cambridge, Mass., 1932–40.
King, Preston (1974), *The ideology of order: a comparative analysis of Jean Bodin and Thomas Hobbes*, 1974.
Koselleck, Reinhart, and Schnur, Roman, eds (1969), *Hobbes-Forschungen*, Berlin 1969.
Krautheim, U. (1977), *Die souveranitätskonzeption in den englischen verfassungskonflikten des 17. jahrhunderts; eine studie zur rezeption der lehre Bodins in England*, Frankfurt-am-Main 1977.
Kretzmann, Norman, Kenny, Anthony, and Pinborg, John (1982), *The Cambridge history of later medieval philosophy*, Cambridge 1982.
Laird, John (1934), *Hobbes*, 1934.
Lake, Peter (1988), *Anglicans and Puritans? Presbyterianism and English conformist thought from Whitgift to Hooker*, 1988.
Letwin, Shirley Robin (1976), 'Hobbes and Christianity', in *Daedalus* 105(1976), 1–21.
Levack, Brian P. (1973), *The Civil Lawyers in England 1603–1641: a political study*, Oxford 1973.
Lund, William R. (1988), 'The historical and 'politicall' origins of civil society: Hobbes on presumption and certainty', in *History of Political Thought* 9(1988) 223–235.
Luscombe, D. E. (1982), 'Natural morality and natural law', in Norman Kretzmann and others, eds, *The Cambridge History of later medieval philosophy*, Cambridge 1982, 705–19.
McLachlan, H. John (1951), *Socinianism in seventeenth-century England*, Oxford 1951.
McNeilly, F. S. (1968), *The anatomy of Leviathan*, 1968.
Macpherson, C. B. (1964), *The political theory of possessive individualism*, paperback edition, Oxford 1964.
Maguire, Mary Hume (1936), 'Attack of the common lawyers on the oath *ex officio* as administered in the ecclesiastical courts in England', in *Essays in history and political theory in honor of Charles Howard McIlwain*, Cambridge, Mass., 1936, 199–229.
Maitland, F. W. (1965), *The letters of Frederic William Maitland*, Cambridge Mass., 1965.
Malcolm, Noel (1981), 'Hobbes, Sandys and the Virginia Company', *Historical Journal* 24(1981), 297–321.
—— (1982), 'Thomas Hobbes and voluntarist theology', unpublished Cambridge University Ph.D. dissertation (no. 12565), 1982.
—— (1984), *De Dominis (1560–1624): Venetian, Anglican, Ecumenist and*

relapsed heretic, 1984.
—— (1988), 'Hobbes and the Royal Society', in G. A. J. Rogers and Alan Ryan, eds, *Perspectives on Thomas Hobbes*, Oxford 1988, 43–66.
—— (1991), 'Hobbes and Spinoza', in J. H. Burns, ed., J. H. Burns, ed, *The Cambridge History of Political Thought, 1450–1700*, Cambridge 1991, 530–57.
Marchant, Ronald A. (1969), *The Church under the Law: justice, administration and discipline in the diocese of York 1560–1640*, Cambridge 1969.
Marshall, John (1985), 'The ecclesiology of the latitude-men 1660–1689: Stillingfleet, Tillotson and Hobbism', in *Journal of Ecclesiastical History* 36(1985), 407–27.
Martineau, C. (1935), 'L'obligation morale peut-elle exister sans la connaissance de Dieu?', *Revue Apologétique*, 60(1935) 258–71, 385–440; 61(1935), 257–76, 401–25.
Mason, A. J. (1914), *The church of England and episcopacy*, Cambridge 1914.
Matthews, A. G. (1948), *Walker revised*, Oxford, 1948.
Milner, Benjamin (1988), 'Hobbes on religion', in *Political Theory* 16(1988), 400–25.
Mintz, Samuel I. (1969), *The hunting of Leviathan*, Cambridge 1969.
Missner, Marshall (1983), 'Skepticism and Hobbes' political philosophy', in *Journal of the History of Ideas* 44(1983), 407–27.
Morrill, John (1980), *The revolt of the provinces*, second edition 1980.
—— ed. (1990), *Oliver Cromwell and the English Revolution*, 1990.
Oakeshott, Michael (1975), *Hobbes on civil association*, Oxford 1975.
O'Day, Rosemary, and Heal, Felicity, eds (1976), *Continuity and Change: personnel and administration of the Church of England 1500–1640*, Leicester 1976.
Okin, Susan Moller (1982), '"The sovereign and his counsellors": Hobbes's reevaluation of Parliament', *Political Theory* 10(1982), 49–75.
Ollard, Richard (1987), *Clarendon and his friends*, 1987.
Orr, Robert (1967), *Reason and Authority: the thought of William Chillingworth*, Oxford 1967.
Pacchi, Arrigo (1965), *Convenzione e ipotesi nella formazione della filosofia naturale di Thomas Hobbes*, Florence 1965.
—— (1988), 'Hobbes and the problem of God', in G. A. J. Rogers and Alan Ryan, eds, *Perspectives on Thomas Hobbes*, Oxford 1988, 171–87.
Packer, John W. (1969), *The transformation of Anglicanism 1643–1660 with special reference to Henry Hammond*, Manchester 1969.
Pagden, Anthony, ed. (1987), *The languages of political theory in early modern Europe*, Cambridge 1987.
Pateman, Carole (1988), *The sexual contract*, Stanford 1988.
Peters, Richard S. (1956), *Hobbes*, Harmondsworth 1956.
Pocock, J.G.A. (1957), *The ancient constitution and the feudal law*, Cambridge 1957.
—— (1973), 'Time, history and eschatology in the thought of Thomas Hobbes', in *Politics, language and time*, paperback edition 1973, 148–201.
Polin, Raymond (1977), *Politique et philosophie chez Thomas Hobbes*, second edition, Paris 1977.

Popkin, Richard H. (1982), 'Hobbes and Skepticism', in Linus J. Thro, ed., *History of philosophy in the making*, Washington 1982, 133–48.
Raphael, D. D. (1977), *Hobbes. Morals and politics*, 1977.
—— (1988), 'Hobbes on Justice' in G. A. J. Rogers and Alan Ryan, eds, *Perspectives on Thomas Hobbes*, Oxford 1988, 153–70.
Reik, Miriam M. (1977), *The golden lands of Thomas Hobbes*, Detroit 1977.
Robertson, George Croom (1905), *Hobbes*, 1905.
Rogers, G. A. J., and Ryan, Alan, eds (1988), *Perspectives on Thomas Hobbes*, Oxford 1988.
Rogow, Arnold (1986), *Thomas Hobbes: radical in the service of reaction*, New York 1986.
Ross, Ralph, Schneider, Herbert W., and Waldman, Theodore, eds (1974), *Thomas Hobbes in his time*, Minneapolis 1974.
Ryan, Alan (1983), 'Hobbes, toleration, and the inner life', in David Miller and Larry Siedentop, eds, *The nature of political theory*, Oxford 1983, 197–218.
—— (1984), *Property and political theory*, Oxford 1984.
—— (1988), 'Hobbes and Individualism', in J. A. G. Rogers and Alan Ryan, eds, *Perspectives on Thomas Hobbes*, Oxford 1988, 81–105.
St Leger, J. (1962), *The "Etiamsi daremus" of Hugo Grotius*, Rome 1962.
Salmon, J. H. M. (1959), *The French religious wars in English political thought*, Oxford 1959.
—— (1991), 'Catholic resistance theory, Ultramontanism, and the royalist response, 1580–1620', in J. H. Burns, ed., *The Cambridge History of Political Thought, 1450–1700*, Cambridge 1991, 219–53.
Sampson, Margaret (1979), '"A question that hath non-*plust* many": the right to private property and the "Engagement controversy", 1648–1652', unpublished University of Sussex M.A. thesis, 1979.
Sanderson, John (1989), *'But the people's creatures': The philosophical basis of the English Civil War*, Manchester 1989.
Schlatter, Richard (1951), *Private property: the history of an idea*, 1951.
Schneider, Herbert W. (1974), 'The piety of Hobbes', in Ralph Ross, Herbert W. Schneider and Theodore Waldman, eds, *Thomas Hobbes in his time*, Minneapolis 1974, 84–101.
Schochet, Gordon J. (1975), *Patriarchalism and Political Thought*, Oxford 1975.
—— (1990), 'Intending (political) obligation: Hobbes and the voluntary basis of society', in Mary G. Dietz, ed., *Thomas Hobbes and political theory*, 1990, 55–73.
Shapin, S., and Schaffer, S. (1985), *Leviathan and the air-pump*, Princeton 1985.
Sharpe, Kevin (1989), *Politics and ideas in early Stuart England: essays and studies*, 1989.
Skinner, Quentin (1965), 'History and Ideology in the English Revolution', *Historical Journal* 8(1965), 151–78.
—— (1966), 'Thomas Hobbes and his disciples in France and England', *Comparative Studies in Society and History* 8(1966), 153–67.
—— (1969), 'Thomas Hobbes and the nature of the early Royal Society', *Historical Journal* 12(1969), 217–39.

—— (1972a), 'Conquest and Consent: Thomas Hobbes and the Engagement controversy', in G. E. Aylmer, ed., *The Interregnum: the quest for settlement 1646–1660*, 1972, 79–98.
—— (1972b), 'The context of Hobbes's theory of political obligation', in Maurice Cranston and Richard S. Peters, eds, *Hobbes and Rousseau: a collection of critical essays*, New York 1972, 109–42.
—— (1978), *The foundations of modern political thought*, 2 vols, Cambridge 1978.
—— (1988), 'Warrender and Skinner on Hobbes: a reply', in *Political Studies* 36(1988), 692–5.
—— (1990), 'Thomas Hobbes on the proper signification of liberty', *Transactions of the Royal Historical Society* 40(1990), 121–51.
—— (1991), 'Thomas Hobbes: Rhetoric and the Construction of Morality', The Dawes Hicks Lecture on Philosophy, in *Proceedings of the British Academy*, 76(1990), 1–61; forthcoming 1991?
Smith, A. Mark (1990), 'Knowing things inside out: the scientific revolution from a medieval perspective', *American Historical Review* 95(1990), 726–44.
Sommerville, Johann P. (1983), 'The Royal Supremacy and episcopacy "jure divino", 1603–1640', in *Journal of Ecclesiastical History* 34(1983), 548–58.
—— (1984), 'John Selden, the law of nature, and the origins of government', *Historical Journal*, 27(1984), 437–47.
—— (1986a), 'History and Theory: The Norman Conquest in Early Stuart Political Thought', *Political Studies* 34(1986), 249–61.
—— (1986b), *Politics and ideology in England 1603–1640*, 1986.
—— (1990), 'Oliver Cromwell and English political thought', in John Morrill, ed., *Oliver Cromwell and the English Revolution*, 1990, 234–58.
—— (1991), 'Absolutism and royalism', in J. H. Burns, ed., *The Cambridge History of political thought, 1450–1700*, Cambridge 1991, 347–73.
Sommerville, M. R. (1984), 'Richard Hooker and his contemporaries on episcopacy: an Elizabethan consensus', in *Journal of Ecclesiastical History* 35(1984), 177–87.
Sorell, Tom (1986), *Hobbes*, 1986.
—— (1988), 'The science in Hobbes's politics', in J. A. G. Rogers and Alan Ryan, eds, *Perspectives on Thomas Hobbes*, Oxford 1988, 67–80.
State, Stephen A. (1985), 'Text and Context: Skinner, Hobbes and theistic natural law', *Historical Journal* 28(1985).
Strauss, Leo (1936), *The political philosophy of Hobbes. Its basis and genesis*, Oxford 1936.
Tarlton, Charles (1978), 'The creation and maintenance of government: a neglected dimension of Hobbes's *Leviathan*', in *Political Studies* 26(1978), 307–27.
Taylor, A. E. (1965), 'The ethical doctrine of Hobbes' in Keith C. Brown, ed., *Hobbes Studies*, Cambridge Mass 1965, 35–55.
Thomas, Keith (1965), 'The social origins of Hobbes's political thought', in Keith C. Brown, ed., *Hobbes Studies*, Cambridge Mass, 1965, 185–236.
—— (1973), *Religion and the Decline of Magic*, paperback edition, Harmondsworth 1973.
Thompson, W. D. J. Cargill (1966), 'Anthony Marten and the Elizabethan

debate on episcopacy', in G. V. Bennett and J. D. Walsh, eds, *Essays in modern English church history*, 1966, 44–75.

Tierney, Brian (1982), *Religion, law, and the growth of constitutional thought 1150–1650*, Cambridge 1982.

Trainor, Brian T. (1988), 'Warrender and Skinner on Hobbes', in *Political Studies* 36(1988), 680–91.

Trentman, John A. (1982), 'Scholasticism in the seventeenth century' in Kretzmann and others, eds, *The Cambridge History of later medieval philosophy*, Cambridge 1982, 818–37.

Trevor-Roper, Hugh (1989), *Catholics, Anglicans and Puritans: seventeenth century essays*, paperback edition 1989.

Tricaud, François (1988), 'Hobbes's conception of the state of nature', in G. A. J. Rogers and Alan Ryan, eds, *Perspectives on Thomas Hobbes*, Oxford 1988, 107–23.

Tuck, Richard (1979), *Natural Rights theories: their origin and development*, Cambridge 1979.

—— (1983), 'Grotius, Carneades and Hobbes', *Grotiana* 4(1983), 43–62.

—— (1987), 'The "modern" theory of natural law', in Anthony Pagden, ed., *The Languages of political theory in early modern Europe*, Cambridge 1987, 99–119.

—— (1988), 'Hobbes and Descartes', in G. A. J. Rogers and Alan Ryan, eds, *Perspectives on Thomas Hobbes*, Oxford 1988, 11–41.

—— (1989), *Hobbes*, Oxford 1989.

—— (1990), 'Hobbes and Locke on toleration', in Mary G. Dietz, ed., *Thomas Hobbes and political theory*, Lawrence, Kansas, 1990, 153–71.

Tully, James (1980), *A discourse on property: John Locke and his adversaries*, Cambridge 1980.

—— (1988), *Meaning and context: Quentin Skinner and his critics*, Princeton 1988.

Van den Enden, H. (1979), 'Thomas Hobbes and the debate on free will', in *Philosophica* 24(1979), 185–216.

Walker, D. P. (1964), *The decline of hell*, Chicago 1964.

Wallace, John M. (1964), 'The Engagement Controversy 1649–1652: An Annotated List of Pamphlets', *Bulletin of the New York Public Library*, 68(1964), 384–405.

—— (1968), *Destiny his choice: the loyalism of Andrew Marvell*, Cambridge 1968.

Warrender, Howard (1957), *The political philosophy of Hobbes*, Oxford 1957.

—— (1965), 'A reply to Mr. Plamenatz', in Keith C. Brown, ed., *Hobbes Studies*, Oxford 1965, 89–100.

—— (1969), 'A postscript on Hobbes and Kant', in Reinhart Koselleck and Roman Schnur, eds, *Hobbes-Forschungen*, Berlin 1969, 153–7.

Watkins, J. W. N. (1973), *Hobbes's system of ideas*, second edition, 1973.

Weber, Kurt (1940), *Lucius Cary second Viscount Falkland*, New York 1940.

Webster, Charles (1975), *The Great Instauration*, 1975.

Wendel, François (1965), *Calvin: the origins and development of his religious thought*, translated by Philip Mairet, paperback edition 1965.

Weston, Corinne Comstock, and Greenberg, Janelle Renfrew (1981), *Subjects and Sovereigns: The Grand Controversy over Legal Sovereignty in Stuart

England, Cambridge 1981.

Weston, Corinne Comstock (1991), 'England: ancient constitution and common law', in J. H. Burns, ed., *The Cambridge History of Political Thought 1450–1700*, Cambridge 1991, 374–411.

Wolin, Sheldon S. (1990), '*Hobbes and the culture of despotism*', in Mary G. Dietz, ed., *Thomas Hobbes and political theory*, Lawrence, Kansas, 1990, 9–36.

Worden, Blair (1984), 'Toleration and the Cromwellian Protectorate', in W. J. Sheils, ed., *Persecution and toleration: Studies in Church History* 21(1984), 199–233.

—— (1985), 'Providence and politics in Cromwellian England', *Past and Present* 109(1985).

Wootton, David (1986), ed., *Divine Right and Democracy*, Harmondsworth 1986.

Zagorin, Perez (1954), *A history of political thought in the English Revolution*, 1954.

—— (1985), 'Clarendon and Hobbes', *Journal of Modern History*, 57(1985), 593–616.

—— (1990), *Ways of Lying: dissimulation, persecution, and conformity in early modern Europe*, Cambridge Mass., 1990.

Zvesper, J. (1985), 'Hobbes' individualistic analysis of the family', *Politics* 5(1985), 28–33.

Index

Abernethie, Thomas, 157
Abiathar, 115, 118
absolute property *see* property, absolute
absolutists, absolutism and absolute power, xiii, 15, 17, 18, 25–7, 36, 52, 58, 61, 63–4, 71, 80, 81, 85, 87, 92–4, 102, 164
  meaning of the terms, vii, 172n35
acception of persons *see* favouritism
Acontius, Jacobus, 147
Adam, 37, 83, 88, 90, 94
Adams, Thomas, 83
adultery, 49–51
Aglionby, George, 11, 21
agriculture, 102, 192n58
allegiance, xii, 6, 7, 36, 67, 69, 70, 114, 115
Almain, Jacques, 34, 35, 132
Amazons, 73
Ambrose, St, 115, 194n15
America, 44
Ames, William, 55, 90, 107, 193n3
Ammianus Marcellinus, 123
anarchy, 35, 37, 54, 59, 83, 99, 108, 126, 131, 154
Andrewes, Lancelot, Bishop of Winchester, 7, 10, 19, 115, 177n13
Andronicus of Rhodes, 44, 45
Anglicans and Anglicanism, 4, 101, 102, 107, 112, 114, 117–122, 124, 127–136, 150, 154–6, 159, 166–7, 196–7n46
  meaning of the terms vii
Answer to the Nineteen Propositions, the, (1642), 185n6
Antichrist, 156
anticlericalism, 105, 169n7
Antioch, 126
Apostles, 111, 113, 120–123, 125, 129, 130, 131, 145, 157, 197n49, 199n11

Aquinas, St Thomas, 34, 47, 48, 53, 90, 93, 137–140, 142, 146, 151, 161, 169n7, 198n9
aristocracy, 16, 81, 83
Aristotle and Aristotelianism, xiii, 8, 14, 15, 28, 40, 44, 47, 55, 137, 157, 159, 161, 169n7, 180n57
  see also scholasticism
Arlington *see* Bennet, Henry
Armada, the Spanish (1588), xii, 5, 128
Arminianism, 8, 136, 148
artificial reason, 98–9, 192n49, 192n50
Ascham, Anthony, 3, 24, 25, 175n72, 175n73
Asia, 44
Athaliah, 115, 118
atheism, xiv, 4, 22, 24–26, 76–8, 137, 141, 166
Aubrey, John, ix, 4–6, 8, 10–12, 18, 21, 27, 81, 168n2, 176n76
Augustinianism, 42
authorisation, 60, 185n6, 186–7n12
Aylesbury, Sir Thomas, 12, 171n26
Ayton, Sir Robert, 11
Azorius, Johannes, 55, 59, 182n73

Bacon, Sir Francis, Baron Verulam and Viscount St Albans, 8, 15, 21, 105, 157, 158, 170n13
Ball, William, 68
Baptists, 105
Barclay, William, 36, 116
Barebones parliament (1653), the, 159
Barlow, Thomas, Bishop of Lincoln, 26
Barlow, William, Bishop of Lincoln, 87, 154
Beale, William, 18, 172n42
bees, 40

222

Index

belief
  knowledge and, 142–3
  obedience and, 109–110
beliefs, fundamental Christian, 136, 146, 148, 149, 152, 159
  see also faith, one necessary article of
Bellarmine, St Robert, Cardinal, 6, 8, 19, 105, 107, 113–118, 130, 157, 159
Bennet, Henry, first Earl of Arlington, 27
Beza, Theodore, 128
bible
  Authorised Version of the, 194n11
  Vulgate, 194n11
  see also Scripture
Biel, Gabriel, 90
bishops
  Anglican views on the powers of, 132, 197n46, 197n48, 197n49
  attacks on Hobbes of the, 26–7
  Cranmer on the powers of, 133
  criticised by Hobbes, 19
  derive jurisdiction from the pope according to Bellarmine, 130
  favouritism and, 48
  Hobbes on the powers of, 23, 113–14, 119–21, 124, 127, 136, 165, 166
  Presbyterian views on, 128
  Presbyterians seen as worse than in 1645, 158
  see also divine right of bishops
Bodin, Jean, 82–86, 91–93, 96, 97, 99, 100, 103, 164
Bolton, Robert, 38
Bordeaux, 174n62
bourgeois values, 180n58
Bowman, Francis, 22
Bramhall, John, 21, 24, 37, 49, 112, 126, 136, 137, 145, 173–4n53, 174n54, 195n22, 195n23
Bridge, William, 34, 35
Buckingham *see* Villiers, George
buggery, 51, 180–1n61
Burgess, Cornelius, 158
Burroughes, Jeremiah, 65

Calvin, John, 42, 69, 109, 128, 138, 142, 143, 199n11
Calvinism, 8, 42, 109, 142, 143
Calvin's case (1608), 69
Cambridge, 8
Camden, William, 180n57
Canne, John, 159
canon law, 53, 60, 177n4
Carneades, 168n7
Carpenter, Nathanael, 178n34
Cartwright, Thomas, 101
Cary, Lucius, Viscount Falkland, 135–6, 159, 166
  Hobbes and, 11
  on punishment, 152–3
  religious views of, 12, 135, 142, 146–50, 153
  Socinianism and, 198n11
Cassander, George, 147
Castro, Alphonsus de, 190n32
casuistry, 4, 53–5, 88, 165
Catholicism and Catholics, Roman, 4–7, 10, 20, 24–5, 37, 53, 59, 87, 99, 101, 102, 105–107, 110, 114, 118, 121, 130, 133, 141, 151, 154, 157, 158, 166
  on church-state relations, 115–17
  on Scripture, 108
Cavalieri, Francesco Bonaventura, 173n45
Cavendish, Sir Charles, the elder, 6
Cavendish, Sir Charles, 6, 12, 13, 21–3, 171n33, 173n45, 174n53
Cavendish, Christiana, Countess of Devonshire, 11
Cavendish, William, first Earl of Devonshire, 6
Cavendish, William, second Earl of Devonshire, xii, 6–11, 170n13, 170n18
Cavendish, William, third Earl of Devonshire, 6, 12, 18–20, 25, 113, 124
Cavendish, William, Earl, Marquis and Duke of Newcastle, xiii, 11–14, 17, 18, 21, 23, 163, 171n33, 173n53
censorship, 24, 105, 166

ceremonies, vii, 131, 141, 151, 154, 155
Chaloner, Thomas, 68
Charles I, xii, xiii, 2, 6, 17, 18, 21, 35, 53, 56, 60, 66, 67, 70, 84–6, 92, 94, 98, 101–2, 104, 124, 151, 163, 175n74
Charles II, xiii, xiv, 6, 11, 21, 23–26, 66, 124, 164, 175–6n74
Charron, Pierre, 45, 147, 168n7, 172n33
Cherbury *see* Herbert
Cheynell, Francis, 109, 142
children, 57
  consent of as basis of parental power, 32, 37, 52, 188n33, 188–9n34
  duties of, 73, 131
  eating of, 184n76
  sacrifice of, 44
  see also family; fathers; mothers
Chillingworth, William, 4, 11, 12, 107–111, 134–136, 142, 146–148, 152, 166, 198n11
Christ, Jesus
  His kingdom not of this world, 117, 120
  His office, 124–6, 129, 195n26, 197n46
church
  of England, 4, 11, 112, 114, 117, 146
  meaning of the term, 117–18, 125–7, 130, 132, 197n49
Cicero, Marcus Tullius, 55, 161
Civil Law, 34, 44, 49, 53, 72, 78, 99, 141, 170n18, 188n32, 189n8
Civil War, the, xiii, 1,3, 6, 12, 17, 20, 27, 34–5, 51, 59, 61, 62, 64–8, 82, 84, 92, 100, 103, 112, 113, 135, 144
  Hobbes' theory and, 30, 84, 162–4, 166
classical republicanism, 3, 169n7, 181–2n71
clergy
  criticisms of the, 23–4, 156–60
  meaning of the term, 194n9
  powers of, 127–34

Clifton, Sir Gervase, xii, 11, 12
coin, power to, 83
Coke, Sir Edward, 27, 45, 69, 96, 98, 99, 169n7, 183n76
Coke, Roger, 45–6, 137, 195n22
Coleman, Thomas, 128, 196n44
Collins, Samuel, 61
colour *see* optics; secondary qualities
Commandments, the Ten *see* decalogue
common good *see* public good
common law, 11, 27, 48, 53, 78, 96, 98, 99, 170n18, 192n51
  see also custom
commutative justice, 47, 48, 180n58
confederacies, 183n76
Connan, François, 53
conquest, 4, 25, 31–3, 56–7, 63–70, 70–1, 73–4, 165, 175n74, 182n71, 184n76, 185–7n12, 187n13
conscience, 53, 65, 109, 135, 149–155
consecration, 119, 122, 133
consent
  as basis of parental authority, 32, 37, 70–4, 188n34
  as basis of political authority, 32, 37, 70–4, 188n34
  conquest and, 25, 32, 63–6, 73–4
  see also contract; covenant; property; taxation
contract, 30–1, 33, 47, 48, 52–58, 64, 65, 72, 78, 96, 98, 164, 181n66
  as basis of political authority, 57, 58, 61, 98, 178n30
  see also consent; covenant
Cook, John, 35, 65, 152
cookery, 116
Corinthians, the first epistle of St Paul to the, 130, 195n26
Cosin, John, 145
Cosin, Richard, 90, 91, 190n32
Cotton, John, 101, 152, 192n54, 194n9
counsel, 47, 96, 109–10, 127, 195n26
courage, 5, 180n58, 185n3
covenant, 31, 48, 51–8, 62, 63–7, 76–78, 90, 93, 94,

95, 104, 164, 165, 182n71,
182–4n76
instituting the sovereign, 57–63;
182n71, 183–4n76
see also consent; contract; fear,
covenants entered into
under; property
Cranford, James, 199n16
Cranmer, Thomas, Archbishop of
Canterbury, 132, 133
Cromwell, Oliver, xiv, 21, 24, 25,
105, 106, 144
Cromwell, Richard, 25
cruelty, 48
Cust, Richard, vii, 170n15
custom, 44–6, 86, 98–100, 192n49
see also common law

Davenant, Sir William, 23
death *see* self-preservation
decalogue, 49–50, 89, 91
De Dominis, Marc' Antonio, 8, 10,
19, 132, 170n13
democracy, 9, 16, 59, 60, 81, 83,
86, 163
deposing power
papal, xii, 5–6, 62, 115–16
popular, 62
Derby, 18
Derbyshire, xii
Descartes, René, xiii, 3, 13–15, 20,
21, 28, 42, 168, 171n26, 171n33,
173n45
despotic government, 63, 71,
87, 92, 94
see also conquest; master; servant
determinism, 21, 137
see also free will
Devonshire *see* Cavendish
Dickinson, William, 70
dictators, 86
Digby, Sir Kenelm, 13, 20, 108, 158,
172n33
Digges, Dudley, 12, 36, 38–9, 43
distributive justice, 47, 48, 180n57,
180n58
see also equity
divine positive law, 54, 97, 101,
195n26

divine right
of bishops, 19, 119, 127, 134,
195n23, 195n24
of kings, 15–16, 25
of tithes, 123
dominion
amongst men, 57
founded in reason not grace, 106
and infallibility, 125
over children, 71–3
over the seas, 13
the pope's temporal, 116
Donne, John, 11, 21, 43–45, 69
Dorislaus, Isaac, 24
drunkenness, 30, 48
Duppa, Brian, Bishop of Salisbury
and Winchester, 26, 176n76
Du Verdus, François de Bonneau,
Seigneur, 22, 174n62

Eachard, John, 2, 195n22
education, 5, 73, 86, 146, 156
Egerton, Sir Thomas, Baron
Ellesmere and Viscount
Brackley, 192n50
Eglionby *see* Aglionby
egoism, 28
elders, 121, 128, 197n47
Eleazar, 118
election
of magistrates, 61, 65
of ministers, 123, 133, 195n25
Elizabeth I, 5, 195n26
Ellesmere *see* Egerton
Elzevir, 20
Engagement, the (1650), xiv, 25, 63,
66–70, 74, 165, 187n20
Enlightenment, the, 169n7
ephors, 82
Epicureanism, 39, 172n33
episcopacy *see* bishops; divine right
of bishops
equality, 29, 40–1, 52, 71, 73, 102,
178n30
equity 48, 100, 103
see also distributive justice
Erasmus, Desiderius, 147, 157
Erastianism, 105, 107–8, 127–8, 135
see also Erastus

Erastus, Thomas, 108, 127–131, 158, 196n34, 196n46
Euclid, 12
Euripides, 5
Eutactus *see* Philodemius
ex officio oath, 52, 181n64
excommunication, 115, 119, 127–133, 196–7n46, 197n49
extempore prayers, criticised by Hobbes, 155

faith
  explicit and implicit, 146
  Hobbes inveighs against those who make new articles of, 153
  Hobbes' views on the nature of, 111, 120, 122, 125, 143, 148
  infallibility of the Catholic church and certainty of, 108, 151
  knowledge and, 142–3
  one necessary article of, 145–7
  reason and, 138, 141–2
  salvation and, 143, 145–8
  Scripture and, 152
  see also belief
Falkland *see* Cary, Lucius
Fall, the
  coercive power and, 39, 41, 178n31
  Hobbes and, 41
  pride and, 83
  property and, 90, 190n27
  salvation and, 42
family, the
  contractualist views on, 58, 71
  Hobbes on, 4, 57, 70–4, 188n27, 188n33
  patriarchalist theory of, 37, 70–1, 188n33
  see also children, fathers, mothers, wives
fathers
  consent, generation, and the powers of, 72–3
  consorting with excommunicate, 130
  Hobbes on the powers of, 32, 57, 70–4, 188n3
  Parker on rights of self-defence against, 35
  patriarchalist theory of the powers of, 70–2, 188n33
  sons may refuse to kill, 104
Fathers of the church, the, 147
favouritism, 48, 95
fear
  covenants entered into under, 31, 39, 55–6, 182n71
  of God, 76–7
  Hobbes' birth and, 5
  of imagined powers, 139–41
  just, 30, 54–6, 165
  soldiers and, 185n3
  in the state of nature, 25, 31, 39
  see also self-preservation
Fermat, Pierre de, 173n45
Ferne, Henry, Bishop of Chester, 35–6, 64, 69
Ferriby, John, 199n16
Fiennes, Nathaniel, 190n17
Fiennes, William, Viscount Saye and Sele, 26
fifth amendment, the, 52
Filmer, Sir Robert, 3, 18, 176n74, 189n2
  his *Patriarcha*, 18, 80, 172n40
  his political ideas, 15, 41, 55, 70–2, 80, 82–3, 85–6, 88, 92, 94, 96, 97, 99–100, 103, 164, 172n35, 188n33
  his views on Hobbes, 3, 18, 77–9, 80
fishing, 192n58
Fitzherbert, Thomas, 118
Fleetwood, Charles, 150
Flood, the, 90
Floyd, John, 65
fool, the problem of the, 182–3n76
Forced Loan, the (1626–7), xii, 9, 50, 56, 80, 84, 104, 170n15
  Hobbes helps to collect, xii, 9, 80, 163
free will, 8, 21, 148
  see also determinism
Fulbecke, William, 53, 55, 191n43

Fuller, Nicholas, 94
fundamental law, 17, 84, 98
Fussner, Frank Smith, 180n57

Gaius, 89
Galileo Galilei, 8, 12, 13, 21, 162
Gassendi, Pierre, 13, 14, 22, 39, 158, 172n33, 179n45
Gauthier, David, 2, 185n6, 185–7n12
Gee, Edward, 3, 79
Gee, John, 158
generation, 71, 72
geometry and the geometrical method, 2, 12, 14, 16, 17, 22, 26, 32, 142, 171n32, 173n45
Gerson, Jean, 37, 47
Gillespie, George, 194n9
God
  the author of nature and the laws of nature, 7, 43, 47, 58, 75–8, 100, 106
  covenanting with, 52–3
  duties towards, 49, 77, 112, 140–1, 155–6
  existence of, 46, 47, 76, 78, 137–40
  Hobbes' views on, 25, 31, 112, 132, 137–41, 155–6, 165–6
  irresistible power of, 75–6, 140
  kings derive their powers only from, 9, 15, 51, 64
  kings accountable only to, 60, 100
  the lord of life, 33
  position of in Hobbes' theory of political obligation, 75–9
  salvation and damnation result from the arbitrary decree of, vii–viii, 8, 143, 148
  the word of *see* Scripture
  see also divine positive law; divine right; law of nature; moral law; providence; revelation; salvation
Golden Rule, the, 31, 78, 153, 177n4
good and evil, 28–9, 163
Goodwin, John, 150–4
grace, 42, 106, 131, 141
  see also faith; revelation
Grégoire, Pierre, 185n1

Gregory of Rimini, 47, 179n50
Grotius, Hugo, xii, 3, 13, 16–17, 44, 63, 79, 97, 109, 147
  political ideas of, 36, 37, 42, 44, 45, 48, 53–4, 55, 58, 66, 72, 90, 94, 96, 115
  religious ideas of, 107, 109, 111, 115, 137–8, 142

Hale, Matthew, 98, 192n49
Hales, John, 12, 146, 147, 198n11
Hammond, Henry, 25, 36, 102, 136, 154, 155, 159, 192n56
Hampden, John, xiii, 17, 21
Hampton, Jean, 2, 177n18, 182n76, 193n60
Harrington, James, 3, 196n41
Harsnet, Samuel, Archbishop of York, 158
Hartlib, Samuel, 13, 105, 158
Harvey, William, 26
Hayward, Sir John, 44, 66, 83, 185n1
heaven, 146–148
hell, 38, 145, 167
Henry VIII, 48, 109
Herbert of Cherbury, Edward, Baron, 13, 138, 157
heresy, 26, 27, 108, 109, 112, 135, 153, 166, 167
hereditary right, 16, 70, 85, 165, 190n12
  see also primogeniture
herile *see* despotic government
Herle, Charles, 61, 62
Heylin, Peter, 18, 82, 189n2
High Commission, 52
High Priests, 115, 118, 145
Hispaniola, 44
Hobbes, Francis, 5
Hobbes, Thomas, the elder, 5
Homer, 27
honesty, 42, 75
honey, 179n45
honour, 59, 73, 86, 95, 104, 139, 140–1, 180n58
Hooker, Richard, 3, 16, 38, 44–46, 88, 109, 117, 132, 199n11
horses, 13
House of Commons, Hobbes'

views attacked in the, 26–7
Hudson, Michael, 35
humanism, 7, 19, 38, 145, 157
Hyde, Edward, Earl of Clarendon, 11, 12, 25, 26, 33, 37, 42, 80, 104, 146, 170n22, 175n70, 175n74, 180n58

idolatry, 44, 101
imposition of hands *see* consecration; ordination
impositions, 93
imprisonment without cause shown, 10
incest, 180n61
incorporeal substances, 25, 105, 140, 156, 166
indefeasible hereditary right *see* hereditary right
Independency, 65, 101, 105, 135, 150, 152, 154
  Hobbes and 26, 114, 121, 195n25, 197n47
indifferent, matters, 141, 154, 155, 167
industry, 102, 192n58
infallibility, 46, 108, 110, 122, 125, 151
infidel *see* pagan
injury, 87, 104
injustice, 4, 43, 48, 66, 87, 88, 92, 94, 95, 100, 104, 165
innocent, punishment of the, 101, 102
inquisition, 155
intemperance, 30, 78, 141
Interdict, the Venetian (1606), xii, 6, 7, 114, 116
Ireton, Henry, 68, 91
Israel, 49, 118, 123, 125

jackdaws, 5
Jackson, Thomas, 71, 88
James I, xii, 6, 7, 9, 11, 15, 27, 36, 63, 69, 81, 83, 84, 96, 103, 114, 115, 123, 192n50
  praised by Hobbes, 102
Jehoiada, 115

Jermyn, Henry, Earl of St Albans, 21
Jerusalem, 126
Jesuits, 10, 37, 48, 55, 65, 71, 99, 108, 118, 146, 151
Jews, 102, 130, 154
Jonson, Ben, 11
Joshua, 118
Josselin, Ralph, 24
judges, duties of, 103
judicial law of Moses, 101
jus zelotarum see zealots, the right of the
just fear *see* fear, just
justice, 14, 42, 43, 47, 48, 50, 54, 66, 121, 152, 178n34, 180n57, 180n58
  see also commutative justice; distributive justice; equity
Justinian, 44, 47, 96, 99

Kellison, Matthew, 99, 115, 192n52
kings, divine right of *see* divine right of kings
knowledge, belief and, 142–144
Kynaston, Sir Francis, 18, 85, 96

language, 9, 40, 47
latitudinarianism, 4, 12
Laud, William, Archbishop of Canterbury, vii, 17–19, 85, 189n2
Laudians and Laudianism, vii, 88, 156
law, human, 96–100, 191n48
law of nations, 44, 191n43, 192n51
law of nature, 7, 15, 28, 30–1, 33–6, 42–51, 54, 55, 72, 74–9, 85, 89–95, 97, 100, 101, 103–4, 106, 107, 116, 141, 152–4, 164–5, 177n6, 188n32, 189n35, 192n51, 195n26
  accessible to all, 78, 153
  God's will and, 31, 46–7, 75–8
lawmaking *see* legislation
Lawson, George, 24
Laymann, Paul, 48, 55, 180n55, 180n58
Le Bret, Cardin, 57
legislation, 61, 84, 93, 96–8

Leibniz, Gottfried Wilhelm, 177n18
Levellers, 3, 62, 91, 152
Levites, 123, 194n9
liberalism, 103, 192–3n58
libertines, 152, 167
liberty, 11, 29, 34, 38–39, 42–3, 83, 88, 150, 151, 181–2n71
  Christian, 113
  Hobbes on, 29, 55, 63, 67, 75, 82, 95, 96, 104, 113, 181–2n71, 186n12
  of the subject, 17, 80, 82, 95, 96, 104
  see also toleration
Lightfoot, John, 196n41
Lipsius, Justus, 169n7
Locke, John, 36, 80, 91, 155, 192n58
London, expansion of, 102
Long Parliament (1640–8), xiii, 18, 27, 80, 81
Lord's Supper, 128, 131
Loudun, the devils of, 158
Love, Christopher, 24
Lucy, William, 143, 188n33
Luke, the Gospel according to, 88, 97
Luke, Sir Samuel, 150

Machiavelli, Niccolò, 169n7
Macpherson, C. B., 180n58
Magdalen Hall, 5
Mair, John, 132
Maitland, Frederic William, 165
Malmesbury, 5
March, John, 68
Marsilius of Padua, 88, 108, 132, 197n46
Marston Moor, xiii, 21
Martel, Thomas de, 22, 174n62,
martyrdom, 125, 149
Mason, Robert (Fellow of St John's College, Cambridge), 8
Mason, Robert (lawyer), 191n38
masters, 63, 66, 88, 97, 130, 131, 186–7n12
  see also servants
materialism, 105, 166, 167
mathematics, xiii, 14, 21, 22, 27, 67, 142

matters indifferent, 154, 155, 200n42
Matthew, the Gospel according to, 97–8, 117, 130, 196n41, 196–7n46
Maynwaring, Roger, xii, 9, 10, 15, 18, 80, 81, 88, 94, 104, 164, 172n35, 172n42, 189n2, 191n38
medicine, 116
Mersenne, Marin, 14, 15, 19–22, 162, 173n50, 198n4
Micanzio, Fulgenzio, 6–8
militia, debate on the (1642), 84
Milton, John, 150, 151
minimalism
  political, 103
  religious, 136, 146, 148, 156, 159
miracles, 141, 143–5, 158, 167, 199n11
Molière, Jean Baptiste Poquelin, 21
Molina, Luis de, 37
monarchy, 16, 26, 57, 60, 64, 81, 83, 87, 94
  as best form of government, 86
  limited, 9, 16, 59, 64
  mixed, 81, 82, 84, 169n7, 185n6
Montagu, Richard, 199n11
Montagu, Walter (Wat), 25
Montaigne, Michel de, 28, 45, 168–9n7, 172n33
Montauban, 174n62
moral law, 7, 49
More, Sir Thomas, 21
Morton, Thomas, of Berwick, 146
Moses, 101, 102, 118, 145
mothers, 5, 71–74, 108, 187n32, 188n32, 188n33
  see also children; family; fathers
Moulin, Lewis Du, 196n42
Moulin, Pierre Du, the elder, 114, 116, 157, 196n42
Moulin, Pierre Du, the younger, 88, 196n42
murder, 35, 38, 49, 50, 192n54

Naaman, 198n1
naked pacts, 53
Naseby, xiii
Nedham, Marchamont, 68

new light, 144, 150, 152, 199n16
Nicholas, Sir Edward, 24, 25, 175n70, 175n74
Noah, 90
nominalism, 2
nuda pacta see naked pacts
Nye, Philip, 152

oath of allegiance (1606), xii, 6, 7, 36, 67, 114–16
oaths, 53, 55, 67, 85, 182n73
  see also ex officio oath; oath of allegiance
obedience, 63, 87–8, 97, 104, 125–6, 143, 148, 149, 154
  belief and, 109–10
  protection and, 66–70
  see also passive obedience; resistance
obligation, 74–9
  see also political obligation
Ockham, William of, 47, 138
omnipotence, 148
optics, 8, 12–14, 20, 21, 170n13, 171n33
  see also secondary qualities
ordination, 119, 120–5, 127, 133, 195n23, 195n25
Osborne, Francis, 68
Owen, David, 36
Owen, John, 26
Oxford, 5, 11, 22, 26, 114

pacts *see* covenants; naked pacts
pagan rites, adopted by Christians, 158
pagan rulers
  duties of Christians under, 126, 127, 149
  powers of, 106
paradiastole, 169n7
paradox, 3, 72
Pareus, David, 36
Parker, Henry, 17, 35, 61, 62
parliament, Hobbes' views on, 96, 191n46
Parsons, Robert, 99, 151, 154
passions, 13, 14, 28, 41, 42, 83

passive obedience, 87, 88, 97, 100, 104
pastors, 115, 118, 121–124, 126, 148
patriarchalism, 71, 73, 94, 188n33
Paul V, xii, 6, 114
Payne, Robert, 12, 13, 23, 24, 124, 127, 171n26, 171n30
peace, 36, 38, 52
  in Hobbes' theory, 1, 29–31, 52, 58–9, 79, 113, 156, 164
Pell, John, 12, 22, 173n45, 174n53
Pemble, William, 192n51
Pentateuch, 145
people, the, 60–2
perfect community, 58, 115
persecution, 147, 150, 153
Peter, St, 117, 118, 130, 132
Petition of Right (1628), xii, 10, 17, 84
Petty, William, 21, 22
Philip II of Spain, 5
Philodemius, Eutactus, 3, 25, 91, 175n73
Phineas, 101, 102
Pilate, Pontius, 109
pills, 106
Pius V, 5
Polin, Raymond, 165
political obligation, 57, 74–9, 189n35
pope, the, xii, 4–8, 10, 59, 82, 114–119, 130, 132, 139, 140, 151, 156, 157, 158, 166
popery, 158, 194n9
popular sovereignty, 36, 60
predestination, 8, 148
  see also Calvinism
prerogative, 11, 18, 80, 83, 85, 170n18, 178n34, 180n57
Presbyterianism, 3, 4, 23, 24, 35, 101, 105, 114, 121, 128, 130, 135, 136, 151, 154, 156–158, 166, 197n47
presbyters, 113, 120, 122, 124
Preston, Thomas, 36, 177n13, 177n14
pride 83, 108, 147
Pride's Purge, 20, 62, 101

Index

priests, 20, 99, 113, 115, 118, 120, 123, 133, 145, 158
primogeniture, 85, 86
private spirit, 108, 110
proclamations, 99, 192n51
procreation, 71, 72
promises, 30, 31, 53, 55, 66, 74, 85, 182n73
propagation, 51, 78, 180–1n61
property, 2, 4, 9, 48–51, 54, 56, 80, 87, 88, 89–95, 104, 163–165, 180n57, 183n76, 191n43, 193n60
  absolute 50, 89, 94
prophecy, 111, 143–5, 167, 199n15
protection, 32, 33, 67–70, 74, 92, 187n20, 188n34
Protestants, 4, 5, 7, 8, 10, 53, 106, 108–110, 114, 119, 127, 144, 145, 151, 157, 158
providence, 16, 70, 75, 101, 110, 144
Prynne, William, 35, 128
psychology, 2, 13, 28, 76
public good, 78, 86, 93, 101–2, 149, 152–3, 162
public interest, 149
Pufendorf, Samuel, 168n7
punishment, 50, 58, 64, 75, 78, 101–2, 135, 139, 152–3, 186–7n12
puritans, 10, 11, 38, 67, 101, 107, 119, 135, 136, 144, 150, 151, 153, 154, 156, 158, 159, 166
Putney debates, the (1647), 68, 91
Pym, John, 172n35, 191n38

Quakers, 26

reason, 45–8, 162–3, 179n48
rebellion, 18, 32, 37, 53, 69, 70, 103, 104, 144, 183n76
religion, 139–41, 178n34
  see also God
representation, 60–1, 115, 132, 145
resistance, 9, 16, 34–37, 50, 51, 66, 76, 87, 100, 125, 177–8n19
Restoration, the (1660), xiv, 4, 26, 27, 145
revelation, 52, 77, 97, 106, 110–112, 124, 137, 141–5, 150, 151, 159, 167
rhetoric, xiii, 2, 168n2, 169n7
Richer, Edmond, 57
right of nature, the, 29–33, 37–40, 43, 49, 52, 58, 59, 79, 95, 186n12
robbers, covenants with, 31, 55
Robertson, George Croom, 17
Roberval, Gilles-Personne de, 173n45
Robinson, Henry, 150
Rogers, Thomas, 109, 130
Roman Law *see* Civil Law
Rouen, 173–4n53
Royal Society, 26
royalism, xiii, 11–12, 18, 20, 23, 35–6, 50–1, 63, 64, 69–70, 71, 82, 88
  Hobbes and, 3, 4, 18, 21, 23–6, 33, 37, 50–1, 52, 59–60, 63–4, 70, 104, 124, 127, 135, 164, 165, 175–6n74
Royston, Richard, 176n74
Rump parliament, the (1648–53), 20, 21, 32, 68, 70, 74, 165
Rutherford, Samuel, 35, 71, 151, 152, 154

sabbath, the, 13, 49, 88
sacraments, 119, 121, 128
  Hobbes' views on, 131
sadlers, 116
St German, Christopher, 53, 181n66
Salic Law, 85
salvation, vii, 8, 42, 78, 129, 136, 145, 146, 147–9, 166, 193n58
  Hobbes' views on, 116, 120, 136, 143, 145, 147–9, 166
Sanchez, Thomas, 53, 55, 181n68
Sanderson, Robert, Bishop of Lincoln, 55, 70, 129, 181n86
Sandys, Edwin, Archbishop of York, 158
Sandys, Sir Edwin, 11, 65
sanhedrin, 196n41
Saravia, Hadrian, 63, 66, 70, 82, 83, 85, 86, 92, 96, 129, 187n13
Sarpi, Paolo, 6–8, 54
Satan, 130, 131

Saul, 118
Saunders, Sir Edward, 48
Saye and Sele *see* Fiennes, William
scepticism, 13, 45, 144, 147,
    168–70n7, 171n30, 172n33,
    179n45
Schochet, Gordon, 188n33
scholasticism, 5, 7, 8, 15, 45, 47, 96,
    99, 105, 156, 157, 162, 172n33
  see also Aristotle and
    Aristotelianism
science, 3, 14, 16, 17, 26, 29, 32, 79,
    143, 157, 171n26, 171–2n33
Sclater, William, 69
Scotus, Duns, 90
Scripture
  Hobbes' use of, 2, 31, 106–7,
    111–13, 115, 117, 118, 131,
    136–7, 143–5, 148–9, 152,
    165, 166
  how known to be the word of
    God, 108–11, 142, 159
  how made law, 97, 110–12, 144
  incompatible with Hobbes'
    ideas, 24
  interpretation of, 112–13, 121,
    122, 124–7, 147, 151, 159–60
  the law of nature and, 7, 16,
    31, 105–7
  which writings are part of,
    108–11, 194n11
Scudamore, John, Viscount
    Scudamore, 18
secondary qualities, 14, 137,
    171n32, 171n33
sedition, 10, 36, 86, 155, 158
  Hobbes' doctrines foment, 37
  Hobbes on, 40, 87, 103, 126, 136,
    149, 155, 156
Selden, John, 11, 13, 39, 44,
    45, 168n7
  Hobbes and, 11, 13, 26, 176n76
  political ideas of, 39, 42–5, 72,
    188n32, 192n56
  religious ideas of, 24, 26, 105,
    123, 128, 176n76
self-accusation, 52
self-defence, 33–37, 41, 50, 52, 58,
    63, 66, 89, 95, 100, 177–8n19

self-interest, 14, 42, 43, 75, 79, 86
self-preservation
  in Hobbes' theory, 3, 15, 29,
    32, 33–7, 39, 40, 43, 44, 46,
    49, 51, 55, 58, 73, 74–9, 103,
    162–3, 164
  in other theories, 33–7, 39, 42, 75,
    168–9n7, 177n6, 178n34
  in Warrender's theory, 75–9
  right or duty?, 76–9
Seneca, Lucius Annaeus, 182n71
servants, 63, 88, 130, 131, 186–7n12
shameful actions, 104
Sharpe, Kevin, 162
Shaw, Samuel, 157
Ship Money, xiii, 17, 50, 56, 100, 104
Short Parliament (1640), xiii,
    172n42
Sibthorp, Robert, xii, 9, 10, 15, 80,
    81, 94, 189n2
Sidney, Algernon, 80, 189n2
singulis major, universis minor, 59, 62
Skinner, Quentin, vii, 1, 77, 168n1,
    169n7, 181–2n71, 187n12,
    187n20
slander, 48
slavery, 63, 65, 66
sociability, 39, 40, 43
Socinianism, 142, 198n11
Socinus, Faustus, 198n11
sodomy *see* buggery
soldiers, 185n3
Solomon, 115, 118
Sorbière, Samuel de, 15, 22, 173n45,
    175n74
soul, 13, 116, 138, 166
sovereignty
  duties of, 100–104
  institution of, 57–63
  powers of, 81–9, 94–100, 119–27
  see also conquest; family; fathers
Sparta, kings of, 82, 83
Spartans, 45
Spelman, Sir John, 82
spiritual *see* temporal and spiritual
    powers
state of nature, 2, 29–31,
    37–43, 49, 51, 52, 54, 56,
    58, 59, 72, 73, 86, 104, 165,

178n20, 178n29, 178n30, 183n76
Stillingfleet, Edward, Bishop of Worcester, 42, 43, 129
Strafford *see* Wentworth
Strauss, Leo, 165, 180n58
Suarez, Francisco, xii, 3, 6, 8, 16–17, 21, 79
  on law, 96–7, 99, 191n48
  on the law of nature, 44–7, 90
  political ideas of, 33–4, 35, 38, 47, 59, 64, 72, 90, 94, 96–7, 99, 102, 115–16
succession, 27, 69–71, 85, 190n12
  see also hereditary right; primogeniture
suicide, 35, 178n34
supernatural, 97, 111, 120, 136, 142, 143
Sweet, John, 146
Symons, Henry, 38

taxation, xiii, 10, 50, 56, 60, 61, 83, 85, 93–5, 100, 103, 104, 163, 167
  see also property
Taylor, A.E., 75, 189n35
Taylor, Jeremy, Bishop of Down and Connor, 12, 39, 166
  political and moral ideas of, 38, 39, 43, 44, 45, 55, 101–2, 191n43
  religious ideas of, 101–2, 129, 147, 150, 153, 155, 158, 166–7
temporal and spiritual powers, 115–121, 125–7
Ten Commandments *see* decalogue
Tenison, Thomas, Archbishop of Canterbury, 39, 42, 51, 195n22
Tew Circle, 4, 11, 134, 136, 147, 153, 159, 166, 167
theft, 49, 51, 89–92
  see also property
Theodosius, 115
Thirty Years' War, the, xii, 7
Thomas, St *see* Aquinas, St Thomas
Thomas, Sir Keith, 180n58, 188n34
Thomson, Richard, 197n49
Thucydides, xii, 9, 11, 27, 163
tithes, 119, 122, 123, 159, 194n9

  see also divine right of tithes
toleration, 134–136, 149–56, 166, 199n27, 200n42
Toricelli, Evangelista, 173n45
Towerson, Gabriel, 72
traditions of the church, 108, 141–2, 147
Trinity, the, 107, 145, 146, 167
Tuck, Richard, vii, x, 168–9n7, 171n23, 171n26, 177–8n19, 193n60
tyranny, 59, 83, 87, 158, 182n71

uniformity, 102, 141, 155, 156, 166, 193n58, 200n42
unity, 60, 121, 155
universities, criticism of the, 8, 26, 105, 136, 157, 159

vainglory, 40–1
Valla, Lorenzo, 38
Vasquez, Gabriel, 192n52
Vaughan, William, 83
Venice, xii, 6–8, 114, 116
villeinage, 94, 191n40
Villiers, Christopher, 198n4
Villiers, George, first Duke of Buckingham, 9
Villiers, George, second Duke of Buckingham, 11, 21
Virginia Company, 11
visions, 143
voluntarism, 179n50
vows, 53

Walker, Clement, 68
Waller, Edmund, 21
Wallis, John, 27, 122
Walwyn, William, 152
war, the state of nature as, 29, 42, 54, 66, 82, 86
  see also Civil War
Ward, Seth, Bishop of Salisbury, 171n33
Warner, Walter, 12, 171n26
Warrender, Howard, 75–9, 188n34, 189n35
wealth, 170n10
Welbeck, 12, 171n26, 171n33

Wentworth, Thomas, Earl of Strafford, xiii, 17
Westminster Assembly, 196n41
White, Thomas, xiii, 20, 138
Whitehall, John, 80
Widdrington, Roger *see* Preston, Thomas
wife *see* wives
Wilkins, John, Bishop of Chester, 171n33
will-worship, 151
William I (the Conqueror), 63, 64

Wilson, Matthew (alias Edward Knott), 108
Wiltshire, 5
wine, 6
Wither, George, 68
wives, 45, 57, 130
women, 1, 6, 72–3, 85, 170n10, 185n3, 195n26
Wood, Anthony, 81

zealots, the right of the, 101, 102, 192n56